MOLTENI
MONDO

MOLTENI MONDO

AN ITALIAN DESIGN STORY

New York · Paris · London · Milan

PHOTOGRAPHED by
JEFF BURTON

CREATIVE DIRECTION by
BEDA ACHERMANN

EDITED by
SPENCER BAILEY

TABLE OF CONTENTS

FOREWORD p.10
Jean Nouvel

INTRODUCTION p.13
Spencer Bailey

THE COMPOUND p.20
Where Molteni&C's Magic Happens

VINCENT VAN DUYSEN p.64
In Conversation with Spencer Bailey

ARCHITECTS AND DESIGNERS p.72
The Maestro Minds Behind the Company

THE ARCHIVE p.148
Molteni&C's Multigenerational Design Legacy

PRODUCTION p.170
Craft and Connoisseurship Inside the Factory

DETAILS p.236
Up Close with Molteni&C's Materials

SHOWROOM p.250
On Set with Molteni&C's Designs

KITCHEN p.258
Three Extraordinary Molteni&C Kitchen Concepts

CONTRACT DIVISION p.272
Molteni&C's Category-Defining
Architectural Collaborations

FLAGSHIP STORES p.296
Molteni&C's Retail Imprint Around the World

MUSEUM p.302
A Curated View of the Company

ROOMSCAPES p.318
Experimental Environments
by Vincent Van Duysen and Ron Gilad

MONUMENTI p.328
The Pillar Pieces of Molteni&C

HISTORY p.354
The Company's Celebrated Trajectory

AFTERWORD p.376
Jacques Herzog

GIO PONTI p.385
Inside the Gio Ponti Archives

FOREWORD

By Jean Nouvel

Every time I've visited Molteni&C's headquarters in Giussano, I've always been impressed – by their precision, by their professionalism, by the way they strive to make things that are absolutely perfect. When Piero Molteni and I first met, I immediately found myself getting along very well with him. He wanted to make my designs even better than I did. It was the first time I had met someone who wanted to make something better than me.

We started working together in the 1990s, when I was designing the Fondation Cartier in Paris and came up with the Less collection. With the first piece, a minimalistic table, I wanted to create something *essential*, in the etymological sense of the word – something that couldn't be simplified any further, almost like the wheel. Over time, I completed the range with different pieces. Alongside the table, there's a storage unit that pivots and opens on both sides – a somewhat enigmatic, totally abstract object. We also made other shelving units, a bench that looks strangely similar to the table, and low coffee tables. The collection caused quite a stir and formed a large part of the identity of the Fondation Cartier at the time of its opening. It has remained a top seller over the years.

The Fondation Cartier is really a 100 percent *designed* building. I had come into making furniture gradually, and at a certain point, I started designing things in conjunction with the buildings I was creating. Clients were asking me for a comprehensive design, so I wanted the furniture to be in keeping with the spirit of the architecture that I was producing. I didn't see why the two – architecture and furniture design – should be separated. I've always been very concerned with that harmony. In my mind, all design is a kind of architecture.

At the Fondation Cartier, the furniture's presence is a part of the architecture and very complementary to it. As things developed, the line went through quite a few transformations, including a change of materials. The tables were made of steel and were therefore very heavy. A table needs to be heavy, whereas a bench doesn't, so the benches became aluminum instead. We also moved into new color palettes, which made it possible to cater to a much wider range of demands.

What's interesting is seeing how these objets d'art can coexist in environments that are completely different from the Fondation Cartier. Often, I'm incredibly surprised to see the contrast between these abstractions with pieces that are in another aesthetic entirely, and yet they remain entirely successful. This isn't a criticism in the slightest. On the contrary, given that they're quite simple – but also quite radical, too – these abstract pieces can fit into many places.

The Less collection happened relatively quickly, and since then, we've had some real adventures as partners – especially at the Musée du Quai Branly in Paris, where I designed all of the scenography. Together, we've managed to create things that went further than if we were separate from one another. All of the work I've done with Piero Molteni – and with the Molteni Group as a whole – is in the pursuit of simplicity of expression in relation to the difficulty of the question being posed. I've always sought out that simplicity as a response to complexity. I consider these furniture pieces to be the quintessence of my work.

INTRODUCTION

By Spencer Bailey

When Giulia Molteni first mentioned the concept for this book, I was instantly intrigued. *Molteni Mondo*, she told me, would feature art direction by Beda Achermann, the Zurich-based former creative director of German *Men's Vogue* who has gained a cult following for his beautiful and exacting work largely in the fashion industry, as well as larger-than-life, Hollywood-style pictures by the Los Angeles-based artist and photographer Jeff Burton, who's widely known for making edgy, lusciously layered images. Clearly, this was not going to be any ordinary book, but rather something filmic, narrative-driven, gorgeously presented, and practically pulled from the glossy pages of a fashion magazine. Molteni&C's headquarters in Giussano, Italy, would quite literally become a film set; the family would serve as the metaphorical film's *"directors"* and its designers would become the *"protagonists."* Script-style, an enchanting story would unfold.

Molteni Mondo was also to be, remarkably, the first-ever book published about the company since its founding in 1934. The stakes were high. I immediately loved the idea, and the fact that this book would be a different and unexpected – and refreshingly entertaining and enlightening – lens through which to view this celebrated Italian design company, its rich history and enduring legacy, and its forward-looking vision as it nears its centenary year. So I joined as the project's editor, and now here I am, writing this book's introduction.

This is where I should note that *Molteni Mondo* may come as a surprise to those who are familiar with the company and its trademark understated elegance. From the start, Molteni&C has produced rigorous, precise, highly technical designs, and over the decades has worked with many of the world's leading designers and architects, including Naoto Fukasawa, Jacques Herzog, Jasper Morrison, Jean Nouvel, Aldo Rossi, Tobia Scarpa, and Patricia Urquiola – as well as Vincent Van Duysen, who has been the company's creative director since 2016. This is a serious business, and the Moltenis are a serious family. But spend a bit of time with them, and a boisterous level of creativity, playfulness, and wonder will quickly unfold. They are an impassioned bunch. *Molteni Mondo* reflects this spirit.

"Mondo" has many meanings that come into play here: In Italian, it translates to *"world,"* with its first usage tracing back to the 1962 Italian documentary film *Mondo Cane*, or *"A Dog's World."* In English, the word has been adopted to mean something striking, significant, or remarkable; in slang, it means *"very large or great in amount or number."* Meanwhile, the Japanese *"mondō"* – etymologically unrelated to the Italian word – refers to a Zen Buddhist phrase for a recorded dialogue designed to obtain an intuitive truth between an old master, or *"rōshi,"* and a student. Even this meaning seems relevant. Each Molteni&C design is, in its own way, a search for its truest essence through material and form.

Across the board, *"mondo"* could be used to describe Molteni&C and the family's history: their journey into design, their obsession with quality and craftsmanship, their respect for production processes, their industrial-scale innovation, their intimate relationships with industry-shaping designers, their operation's ongoing growth. From its rise in the 1950s and '60s creating truly modern furniture pieces – including Werner Blaser's MHC.1 dresser prototype (1955) and Yasuhiko Itoh's MHC.2 bookcase (1959), and breakthrough systems by Tito Agnoli, Angelo Mangiarotti, and Luca Meda – to the 1990s, when it debuted the category-defining Less collection by Jean Nouvel, to today, Molteni&C has been a central force within the global design conversation. More recently, the company's profile has risen and been further transformed under the direction of Van Duysen, who has brought his pared-down Belgian imprint and minimalist rigor to the company's many surfaces, alchemizing these sensibilities with his deep understanding of and passion for Italian design.

Molteni&C's sprawling, 1.6-million-square-foot (150,000 square-meter) Compound in Giussano is its own *"mondo,"* too, home to the company's factory, R&D lab, and offices, as well as its archives; the Urquiola-designed Spazio QallaM *"designer space,"* which hosts seminars, workshops, exhibitions, and installations; the Ron Gilad-designed Molteni Museum, which presents a permanent collection of more than 50 products; and the campus's latest addition, the Van Duysen-designed Pavilion, which includes a lobby, a restaurant, and an outdoor garden. Just beyond the Pavilion, an original building from the 1950s features a three-floor showroom inside.

      The Molteni&C headquarters feels not unlike being on the Hollywood lot of Paramount Pictures: there's something cinematic and lights-camera-action about it. In constant motion, with company president Carlo Molteni whirring around the grounds on his cherry-red bicycle, the place is an impressive sight to behold. Between the factory and offices, roughly 600 employees make up this beautifully choreographed reality, with all of the different departments and zones – fabrics here, leather there; cutting here, stitching there; lacquering here, shipping there – coming together in profound unity, like the gears of a clock.

      I've visited many factories around the world – other furniture-makers in Europe, metal fabricators in America, high-end watch studios in Japan – and I've never seen an operation quite like Molteni&C's. During a tour of the Compound, I was able to watch as the family met with Salvatore Licitra, the head of the Gio Ponti Archives and grandson of the renowned designer, to go over every last detail of a Ponti sofa design they were planning to reproduce, based on a single archival image alone. A nearly hour-long conversation ensued, with an animated discussion between them about each and every angle, corner, and curve. In today's increasingly conglomerated design world, Molteni&C stands out not only as a family-owned operation but also for its *"mondo"* approach. This is a company that puts its designers and clientele first, rigorously realizing a vast range of impeccable products that, as soon as they leave the walls of Giussano, go on to resonate and make global impact.

And … action!

THE STORY BEGINS...

MOLTENI

SCENE SLATE

THE CON

ROLL

DIR JEFF B
DOP
DATE

The MOLTENI FAMILY and CEO MARCO PISCITELLI on set in the Compound, Giussano, Brianza, Italy

The MOLTENI FAMILY and MARCO PISCITELLI get comfortable

Carlo Molteni's cherry-red Olmo bicycle

Carlo Molteni takes a ride on his bike at the Molteni&C compound

Where Molteni & C furniture leaves the factory and enters the world

The <u>BRICK TOWERS</u> of the industrial plant at the compound, symbols of the company's circular economy

Bottom left and top right: Spazio QallaM, designed by Patricia Urquiola in 2006

Even the Molteni & C trucks exude an understated elegance

The Pavilion, designed by VINCENT VAN DUYSEN and opened in 2022

The Pavilion's lobby entrance and front desk

The Pavilion's RESTAURANT

THE COMPOUND

By Maria Cristina Didero

Representative of Molteni&C's ongoing growth and strong position on the international design stage, the company's Compound in Giussano, Italy, just forty minutes by car from Milan, serves as a link between — and an embodiment of — its storied past and bold vision for the future. The original layout of the site dates to the early 1950s and is proudly preserved via certain enduring architectural elements that have stood the test of time, including a Pierluigi Cerri staircase, timeless facade cladding, and characteristic window frames. Central to this legacy is a four-story structure designed by Luca Meda, with a round pillar bearing the distinctive architectural signature of Aldo Rossi. Standing as a proud testament to the company's unwavering commitment to its design heritage, the structure underscores the profound influence that Rossi had on Molteni&C.

Bringing this cherished legacy of Italian design and innovation into the twenty-first century, Molteni&C unveiled the Pavilion in 2022. Designed by the Belgian architect and designer Vincent Van Duysen, the company's creative director since 2016, the ambitious expansion of the existing site spans an impressive 15,069 square feet (1,400 square meters). As described by Carlo Molteni, president of the Molteni Group, the Pavilion is a contemporary edifice seamlessly integrated into the historical site, symbolizing the company's forward-looking trajectory. Marking a pivotal moment for Molteni&C, the Pavilion represents a pronounced architectural expression of the company today. Van Duysen's vision revolves around the idea of creating an ongoing dialogue between interior and exterior spaces, with the structure's floor-to-ceiling windows opening up to an interior courtyard. A continuous connection between the building's interiors and its surroundings redefines the traditional boundaries between indoor and outdoor spaces; from the showroom's third floor, visitors can even see the Dolomites far off to the north.

To bring his concept to life, Van Duysen first embarked on the redesign of the guest entrance, introducing a concrete walkway enveloped by lush greenery and a sleek gatehouse. This transformation offers visitors a captivating preview of the colonnaded corridor, which welcomes them as they walk along a picturesque garden to the Pavilion's reception area and restaurant – the new heart of the Compound. Characterized by its low, modernist-inspired structure, the Pavilion features expansive windows that can be opened to let in light and fresh air. The renovation also reimagines the two cloisters of the Compound as al-fresco exhibition spaces that house select pieces from the Molteni&C outdoor collection. These elements meld with the sinuous, plant-filled pathways artfully crafted by landscape architect Marco Bay. The courtyards serve as environments, in Bay's words, *"where the natural world is allowed to express itself through organic forms, varying heights, captivating textures, and dynamic elements, [and] where the passage of time will organically nurture an unrefined tapestry of greenery, enfolding the furniture components and inspiring fresh décor concepts."*

Taking cues from its surrounding columns, the welcoming space of the Pavilion functions as a versatile structure. Nestled within the cloisters, it seamlessly integrates the reception area, with the expansive Modernist-influenced glass facades creating an immersive experience for visitors. The restaurant, meanwhile, offers a hospitality space for dining and conviviality, with picturesque garden views. The design, which combines glass and concrete with light oak walls, represents another successful collaboration with Van Duysen, who's known for his focus on simplicity and warmth, as well as for his meticulous attention to detail. The ground floor of the showroom is now known as *"Sala Luca Meda,"* honoring the namesake designer's vital role in transforming Molteni&C from a traditional furniture manufacturer into a globally recognized design company. The Sala features boiserie walls and an elm-colored ceiling that starkly contrast with the exterior colonnade's darker tones.

Nearby, just west of the factory, is the glass-cube Molteni Museum, which boasts an elegantly arranged 4,306-square-foot (400-square-meter) exhibition space flooded with natural light. Inaugurated in 2015, the Museum was established to celebrate the company's history of innovation, research, and commitment to quality. In 2021, it underwent a comprehensive redesign by the Israeli-born designer Ron Gilad, the museum's curator and a longtime collaborator of the company. A dynamic and sophisticated design that illustrates Molteni&C's rich narrative, this space offers an unconventional perspective on the company's archive, showcasing its life, products, prototypes, graphics, and imagery all while spotlighting its core identity and values.

Adjacent to Gilad's Museum is Spazio QallaM, from 2006, a multimedia space designed by the Spanish-born, Milan-based architect and industrial designer Patricia Urquiola, another Molteni&C collaborator. Inspired by her 2005 Diamond table for Molteni&C, Urquiola created an 82-foot (25-meter) promenade leading to a panoramic room at the top. This room, often used for film projections, offers views of the surrounding park and features a glass wall adorned with geometric motifs.

A dynamic hub and a true laboratory for the future, the Molteni&C Compound transcends the role of a mere production and business venue. Beyond being a mere showroom for displaying Molteni&C's latest designs or simply a production facility, the Compound offers a comprehensive narrative experience for visitors, serving as a portal into the company's rich history. This headquarters exemplifies a *seamless* fusion of architecture and furniture, delivering a holistic experience that showcases Molteni&C's unwavering commitment to design excellence and to fostering a distinct company culture. This dedication has been a hallmark of Molteni&C from its very inception, and as the Pavilion indicates, it continues to flourish.

The R&D TEAM and Molteni&C's Creative director, Vincent Van Duysen, check prototypes of the LOUISA tables collection

Van Duysen checks the details of the ASTER table top

To make each piece of glass, a sheet is placed in a mold with a rusticated bottom and heated at the temperature of 180° Fahrenheit

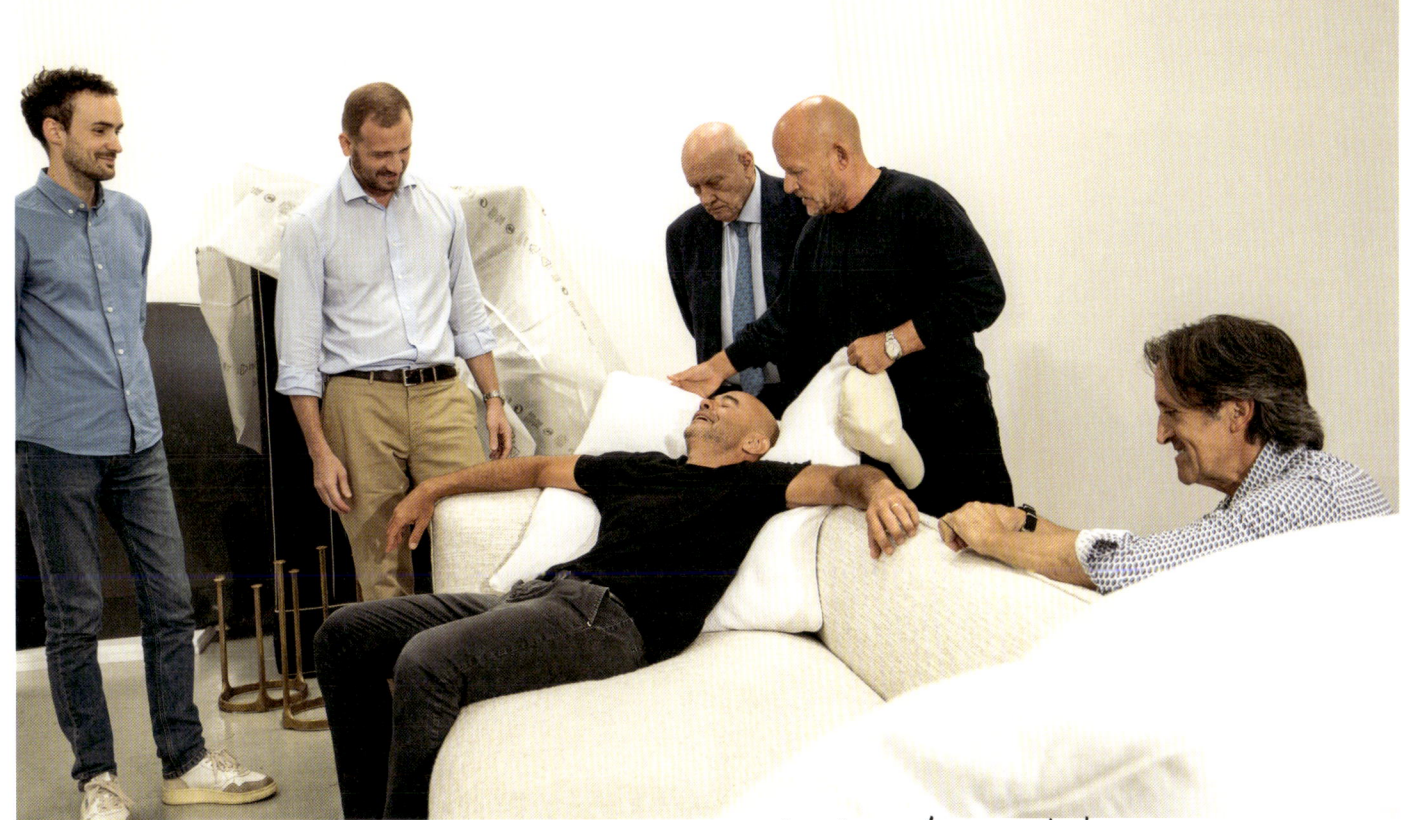

Van Duysen and the R&D TEAM test sofa prototypes

Prototypes of new furniture

from left to right: PAULA armchair and ottoman, LOUISA coffee tables, BREEZE sofa

"WE WANT TO B[E]
DRIVEN AND TO T[HINK]
THAT PEOPLE C[OMPRISE]
THE COMPANY, A[ND IN A]
'MONDO' SEN[SE]
I'M CONTINUOUSL[Y]
AND TOWARD

E NARRATIVE –
ELL STORIES SO
AN UNDERSTAND
S WELL AS THIS
SIBILITY.
 WORKING WITH
HAT 'MONDO.'"

VINCENT VAN DUYSEN

Vincent Van Duysen and Nicola Gallizia select fabrics for new projects with the R&D TEAM

Catalog images featuring Molteni & C's PAUL sofa in four different configurations

The Pavilion courtyard

VINCENT VAN DUYSEN

In Conversation with
Spencer Bailey

SB: What was it like to be named the creative director of Molteni&C?

VVD: For me, it was an incredible honor. When I started in this role, I was a bit shy and slightly insecure. I knew about the rich history and the DNA of the brand. I'd been a great admirer of Molteni&C ever since I was an architecture student [at the Higher Institute of Architecture Sint-Lucas, Ghent, in the early 1980s]. Back then, I wasn't even 20, and went to Milan because I was very Italian-minded. I remember going to the Salone del Mobile, where the Molteni&C booth was outstanding. It was when they were working with Aldo Rossi. I was a kid of postmodernism, so seeing it was a big deal to me. Then, in the 1990s, Jean Nouvel came in. For me, this was, *"Wow!"* Rossi, Tobia Scarpa, Gio Ponti, Jean Nouvel–this heritage was really breathtaking to me as a young architect.

SB: How and when did Molteni&C approach you about the role?

VVD: They had been following my work for a while through publications. My earlier residential work was–and still is, I hope–distinctive. When we met, I immediately got along very well with Carlo Molteni. He's someone who prefers to communicate with an architect who understands what we call *"the art of living"*–not only a sense of space, but also whatever belongs to that space, how you can make spaces come alive, the interaction with light and other sensory elements. From the beginning of my career in the early '90s, I paid a lot of attention to this. There was always a warm, domestic feeling to my work.

These elements–this *interconnection*–is something I've always taken with me, and always in a very subtle, not-too-noisy, and understated yet elegant way. People think of me as Belgian, as someone who creates spaces with these desaturated color palettes, mainly with natural materials, stones, woods, beautiful fabrics and carpets. But I also studied and worked in Milan [collaborating with Aldo Cibic at the studio Sottsass Associati between 1986 and 1987] and traveled a lot, so my architectural expression comes from that mix, too.

At a certain point, the Molteni family reached out to me to collaborate. I'm a very selective person, and if I work with a company – and also in my private residential work, with a client – the primary reason is always the chemistry between us. There has to be a sense of empathy. There has to be mutual respect, and there has to be a lot of humanity. One day, the family came to visit me in Antwerp at the studio and my house. They saw the way I live, and they all immediately fell in love with my home – the warmth, the proportions. For them, it's essential that things be comfortable, approachable, and livable. The *"handwriting"* of an architect, I think, was something very important for them, specifically for Carlo, because he has always worked closely with architects – Aldo Rossi, Tobia Scarpa, Jean Nouvel. Luca Meda was the creative director before me.

In the beginning, they asked me if I wanted to do something around their Gliss Master [walk-in closet system]. They understood that many of my interiors were an extension of my exteriors and interwoven with the world outside. They liked how I dealt with, let's call it, *"domestic architecture."* So for Gliss Master, we created some new panels, new décors, new hinges and door handles. They also asked me to design a bed. The first bed was very minimal, with beautiful details – with fabric, with an inlay in leather on the corner to give a sense of shade and depth.

Next they asked me whether I was interested in designing their booth for the IMM Furniture Fair in Cologne. Because they were impressed by my house – the proportions, the colors, the textures, the materials, the patios, the hidden secret garden – I thought, Okay, I'm going to take that with me as a starting point. I went back to these archetypal forms – wooden floors, ceiling beams, steel window frames – in a very artisanal way. It was a space seen through me, but with all the elements and essences that I took from Molteni&C. Because they'd long had this very linear and graphical and geometrical design approach, I thought, I'm going to continue in that line, but I'm going to lend the Vincent Van Duysen warmth to it in terms of materials and colors.

This new concept was a huge success. The family loved it. Everybody loved it. I loved it, as well. I was very surprised. It was all about mass and void, inside and outside, the layout of the furniture, all these layers. In the booth, people could imagine how they could live with these pieces in their own homes.

    SB: Right. It wasn't a *"booth"* necessarily–it was more like a case-study house.

    VVD: Exactly, it was like a pavilion. A week or so later, the family said, *"Vincent, we're so happy with what you've done with us. Would you be interested in becoming our creative director?"* I was around 54 years old. Now I'm 61, so it's been seven or eight years.

I'd already hit certain milestones in my career, but for me, this was a cherry on top of the cake. Being a Belgian architect, becoming creative director of one of the major leading companies in design worldwide, this was huge. Before I started working with them, I had been invited to Molteni&C's eightieth-anniversary celebration at the Galleria di Arte Moderna in Milan. Then, before I knew it, I was their creative director.

    SB: How does it feel to be in this role now? What's your vision for it?

    VVD: It's a huge challenge and responsibility. Psychologically, at first I thought, What if it doesn't work out? But, of course, there had already been a tryout with some positive results. I'm a very calm person, so I said, *"Okay, let's do it."*

I don't have to tell you how driven and passionate the Molteni family are and I am. The whole family are now my friends. The company is now my family. I'm so grateful. They basically gave me carte blanche, and the whole world that I've created around them has been done in a very consistent way, always reflecting a pure sense of living and inherited inspiration. Ultimately, Molteni&C is an Italian company. I don't want to get away from that. It's a beautiful company where you can tell it's all crafted and made in Italy. There is an incredible heritage and legacy that we're taking with us.

    SB: What has it been like for you to go into the company's archives?

VVD: I don't want to look too literally at what's been done in the past. I'm a very intuitive visual person. I'm like a sponge. It's very easy for me to see how things are linked with Molteni&C, and to reinterpret them in my own way, not going too deep and analyzing the designs too much, but just vaguely taking it up in myself and giving the pieces a kind of Italian-istic touch.

As a Belgian, I'm feet-on-the-ground, very humble, very passionate. I really want to go further with them, to make sure that we reach out as far as we can and inspire the world. I always say that with our beautiful pieces there is an *"understated elegance."* Sometimes, I don't mind giving touches of eccentricities, but I'm someone who has always stayed away from trends. I'm a modernist at heart, and carry with me my classical and conservative values as well.

My work is very tactile, very textured, very sensorial. These elements are something that I've brought with me in my journey with Molteni&C–and in a very broad spectrum, not only the furniture pieces, but contextually, through storytelling, advertising campaigns, images, and films. We want to be narrative-driven and to tell stories so that people can understand the company. Not only see the company, but also *understand* this *"mondo"* sensibility. I'm continuously working with and toward that *"mondo."*

SB: Part of this effort is your design of the Pavilion, which serves as both a literal entrance to the headquarters in Giussano and a metaphorical entrance to this *"mondo"* experience.

VVD: The Pavilion is the core. It's the epicenter of the premises. Every time I'm at the company headquarters, I'm so amazed by this mixture of industrial buildings that the family built. At some point, I told them, *"We have to give more identity to where we are now, in the twenty-first century, and to what Brianza means, and to what Molteni means, already with these very strong architectural interventions."*

As Molteni&C's creative director, I wanted to contribute to the heart of the premises. I asked, "What if we create a pavilion that will host people, where we welcome our guests, where people can approach the company via private parking and a walkway in an arcade, in a sequence?" Architecturally, it adds a twist to the company and is now part of what Molteni&C is.

> SB: How do you view yourself within the company's nearly century-long history?

VVD: That's a tough one. The two of us, Vincent Van Duysen and Molteni&C, are here to reach out to the world, and to make sure that we're contributing to people's quality of life. That's what it's all about, passing on this message globally: that as an industrial brand, there's a sense of well-being to what we do, not just design for the sake of design. That's my main goal through working with Molteni&C: to contribute to a high quality of life, and to export the brand into all of these cultures all over the world.

> SB: This gets back to the *"mondo"* idea – it literally means *"world."*

VVD: I think it's so important that we bring that soul and emotion to what we're doing and creating. This is how the Molteni family thinks, too. I do this with the eye of an architect, but I do it also with the eye of someone who knows how people live in their homes.

As a young architect, I started my career diving into interiors because I wanted to know how people were living. I wanted to know what people were asking from their architects in terms of, How can an architect contribute to my quality of life? As a young architect, you're not necessarily trained on that. You're trained purely mathematically about architecture, and in a very rigid, conceptual way. I wanted to know how a house functions from the moment that you approach it. You open the door, then it's a circuit of routines until you go to bed, wake up, and go out again. I still think about all of this today. It's something that I bring to my work with Molteni&C.

> SB: This approach must lend itself well to the intimate, family-run nature of Molteni&C and working with its management team.

VVD: That's the glue. That's the beauty of it. Also, I'm one of the rare architects who, even though I'm involved in so many private residential projects, is still so confident. A lot of other architects don't want to deal with private clients because it's so intense. You almost have to be a psychologist. But I love that aspect of it. Clients inspire me, as I inspire them. This is what I take with me: to try to understand what they're asking for and what they're expecting from me. Whether I'm creating a house or creating something together with the Molteni family, this is how it works.

VINCENT VAN DUYSEN In the "goods" lift with the PAUL sofa

FOSTER + PARTNERS designer MIKE HOLLAND

Scopes out the ARC table (2010)

Sketches, 3D models, and prototypes of the ARC table

Patricia Urquiola

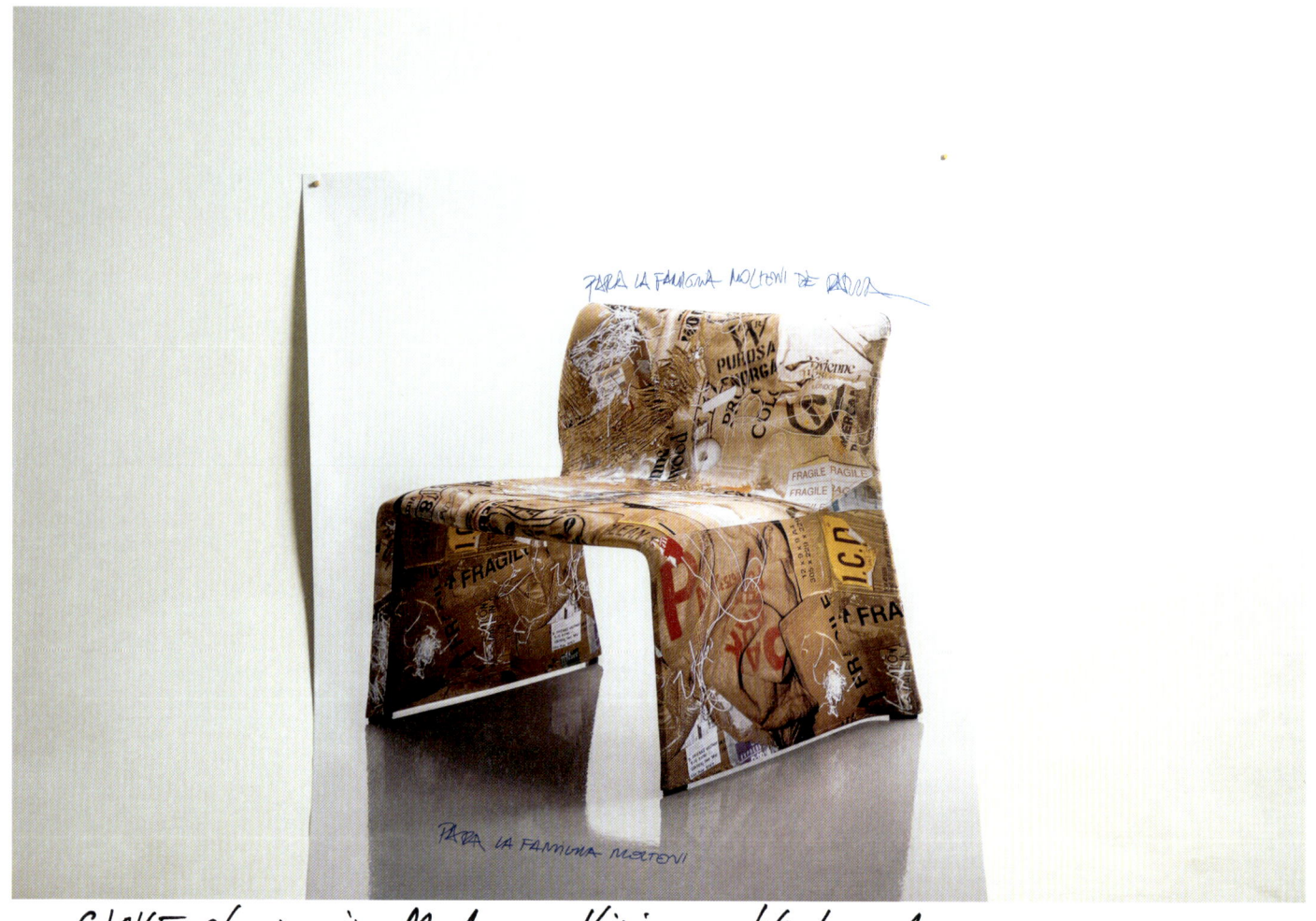

GLOVE chair (2005) by Patricia Urquiola

GLOVE chair in Madame Vivienne Westwood fabric by Vivienne Westwood (2007)

Urquiola looking out from her <u>QallaM</u> "designer space"

A leg of the DIAMOND table (2004) designed by Urquiola and the inspiration for <u>QallaM</u>

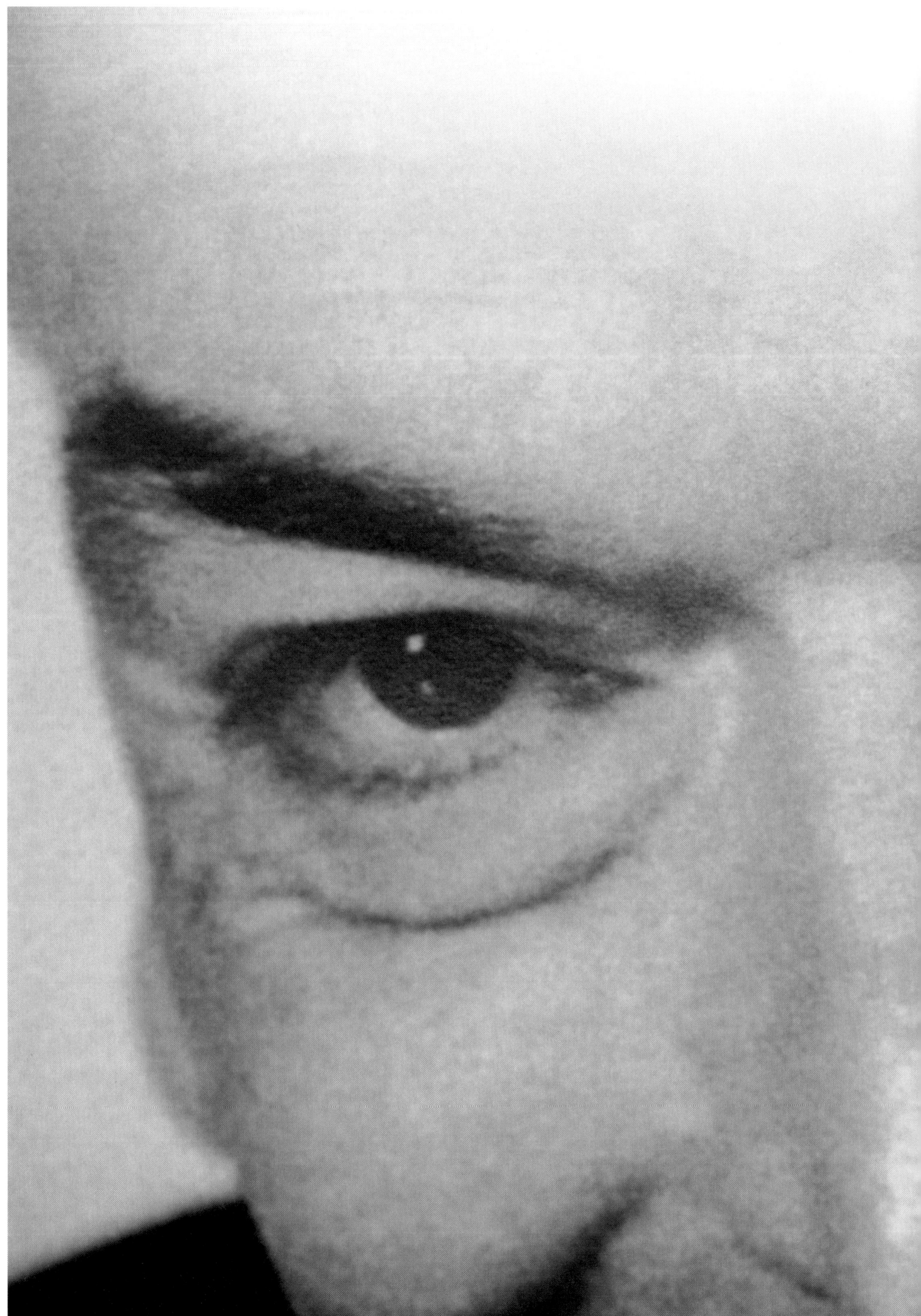

JEAN NOUVEL

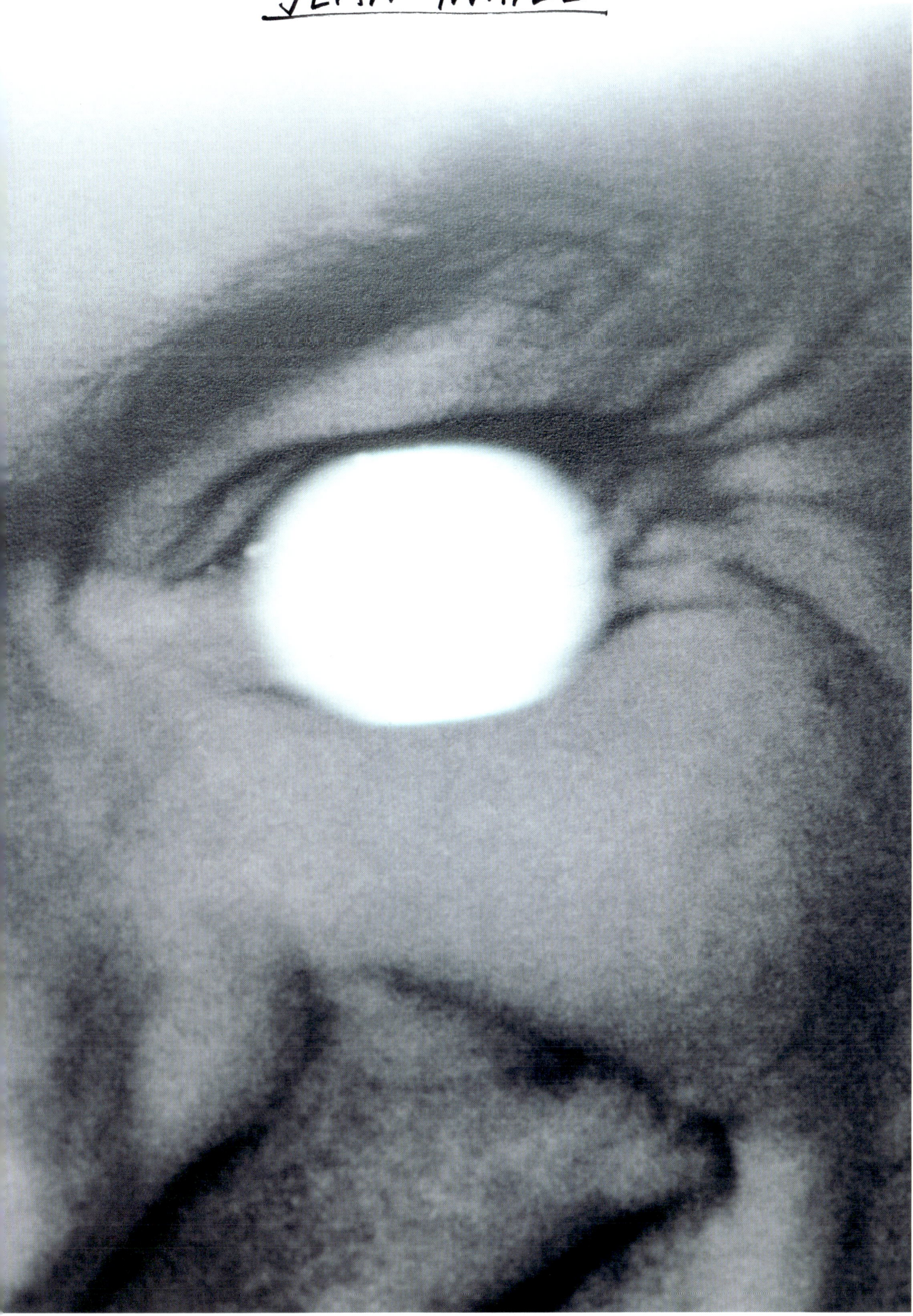

GRADUATE bookcase (2003) and LESS table (1994), both by Jean Nouvel

Jacques Herzog

In his studio in Basel, Switzerland

Models of the PORTA VOLTA chair (2023) by Herzog & de Meuron

MIKE HOLLAND
(FOSTER + PARTNERS)

As head of industrial design at the international architecture firm Foster + Partners, founded by the renowned British architect Norman Foster in 1967, Mike Holland is at the helm of products designed for large-scale projects such as airports, offices, and hotels. Each of their decisions has a ripple effect on the environment and on the daily well-being of many.

Born in the U.K., Holland has been fascinated since childhood by the ways things come apart and are put back together. It is this inquisitive spirit that has underpinned his career as a designer. He studied furniture and product design at Ravensbourne College in London before joining Foster + Partners in 1995. Whether industrial design solutions for the National Bank of Kuwait or the Bloomberg building in London, a fountain pen or a door handle, a wind turbine or a yacht interior, each of Foster + Partners' ideas begins its life at the firm's model-making shop. From there, specialist engineers, artisans, and manufacturers such as Molteni&C, Louis Poulsen, or Vitra join the team in order to build a prototype. It's at this collaborative stage where Holland believes design gets truly interesting.

With Molteni&C, Foster + Partners has created the Still sofa (2006), the Arc table (2010), the Teso coffee table (2013), the Ava table (2019). Streamlined, minimal, and classic in form, and often inspired by structural engineering, these designs demonstrate the industrial design team's decisive mastery over materials—and the ability to find their strength and release it in its purest, most economical form.   —Harriet Thorpe

PATRICIA URQUIOLA

An innovator focused on user experience, Patricia Urquiola (born in Oviedo, Spain, in 1961) merges humanistic, technological, and social approaches through her architecture and design while also tackling issues such as mobility and sustainability. In her reimagination of the world, she connects craftsmanship with technology, heritage with future-forward visions, and reused materials with novel ideas.

Urquiola blends Spanish and Italian design influences, informed over her career by Italian mentors including Achille Castiglioni, Maddalena De Padova, and Vico Magistretti. She studied architecture and design at the Technical University of Madrid, then at the Polytechnic University of Milan under Castiglioni — to whom, upon graduating in 1989, she became assistant lecturer. Later, she joined De Padova, where she was responsible for the R&D office and collaborated with Vico Magistretti on various products.

In 2001, Urquiola established her studio and today leads an international, multidisciplinary team of 70. Her projects with Molteni&C exhibit her ability to artfully juxtapose qualities of classic and contemporary thinking, from her Diamond tables (2004), which exude sculptural lightness through folded aluminum; to her Asterias tables (2016), which employ 3-D printing for their bases, inspired by the form of their namesake Mexican cactus; to her Scriba desk (2010), which comprises a clean-lined workspace with a perforated, decorative top.

Urquiola has won several international design awards and plays a leading role on the global design stage today. She is on the advisory board of both the Polytechnic University of Milan and the Triennale Milano Museum, and travels the world teaching and lecturing at institutions such as Harvard University, the Vitra Design Museum, and Design Shanghai. —H.T.

JEAN NOUVEL

The architect Jean Nouvel (born in 1945 in Fumel, France) approaches every building he designs—from the Musée du Quai Branly in Paris (2006), to the Louvre Abu Dhabi (2017), to the National Museum of Qatar in Doha (2019), to the Museum of Art Pudong in Shanghai (2021)—with a fresh vision guided by both the present moment and the site's history and environment. After graduating from the Beaux-Arts de Paris, he worked under the architect Claude Parent (1923-2016) before establishing his own studio in 1970. Nouvel was awarded the Aga Khan Prize in 1989 for his first major building, the Arab World Institute in Paris, and numerous awards have followed: the Golden Lion of the Venice Biennale (2000), the RIBA Royal Gold Medal (2001), the Praemium Imperiale (2001), the Borromini Prize (2001), and the Pritzker Prize (2008).

As an architect who also designs furniture, he embarked on his first furniture projects in 1987 in the context of VIA's *"Carte Blanche"* series. Later, he developed several collections related to his architectural projects. In 1995, he expanded his practice by founding the company Jean Nouvel Design (JND). Since then, JND has developed more than 100 products with leading design companies, including Alessi, Artemide, and Molteni&C.

Molteni&C has been an important collaborator on Nouvel's interior and design work since the 1990s. The Less collection (1994), which was followed by Less-Less (2012), includes a minimalistic aluminum table, shelving and storage units, and an aluminum bench, all hyper-refined to the point of abstraction—forming the identity of his architecture for the Fondation Cartier in Paris. Nouvel's deep relationship with Molteni&C has also led to the Graduate modular shelving and storage structures (2003), made of aluminum-coated plywood with concealed steel braces to achieve lightness, as well as the Skin sofa (2007), comprising a self-supporting hide leather cover perforated with a pattern. As with everything Jean Nouvel does, these designs are engineered for simplicity and visual lightness. —H.T.

HERZOG & DE MEURON

The buildings of the Basel-based architecture firm Herzog & de Meuron represent the cutting-edge forefront of twenty-first-century architecture. Founded by Jacques Herzog and Pierre de Meuron (both born in 1950 in Switzerland), the powerhouse practice and its staff of more than 550 is behind the designs of some 600 projects in 40 countries – from opera houses and hospitals to libraries and stadiums. In 2001, the pair was awarded the Pritzker Prize.

H&dM is known for designing evocative buildings that are expressive of their locations and depart from expectations. Their highly engineered structures respond to the demands of contemporary urban life and make statements in the process: just like how their architecture can define the identity of a city (consider the Elbphilharmonie in Hamburg, from 2016, or the Tate Modern in London, completed in 2000 and expanded in 2016), H&dM's designs can shift how one perceives space.

Under H&dM Objects, the practice collaborates with manufacturers and craftspeople to design furniture, lighting, textiles, and interior details often specially created for their architectural projects – such as furniture for the Kinderspital Zürich (2024) or a cork stool inspired by their Serpentine Pavilion (2012) – yet also connected to the founding architects' lifelong interest in material experimentation. It's a quality found in the Porta Volta chair (2023), designed for Molteni&C by Jacques Herzog, who worked closely with the craftspeople of the H&dM Atelier to achieve a grounding and sensual form in solid wood with an upholstered seat pad and deep backrest, available in materials including mohair velvet and in colors such as sunflower yellow and royal blue. Featured in the main reading room of the National Library of Israel in Jerusalem, it is equal parts elegant and relaxed. —H.T.

MARTA FERRI

With the fabric collection she curated for Molteni&C

RODOLFO DORDONI (1954-2023)

Pieces from his CHELSEA armchair and sofa collection (2015)

STUDIO KLASS — Alessio Roscini

With the TOUCH DOWN UNIT workdesk (2020)

Marco Maturo

JASPER MORRISON

Patricia Urquiola's GLOVE-UP chair (2016) with a Molteni&C INK writing desk (2016), by Jasper Morrison

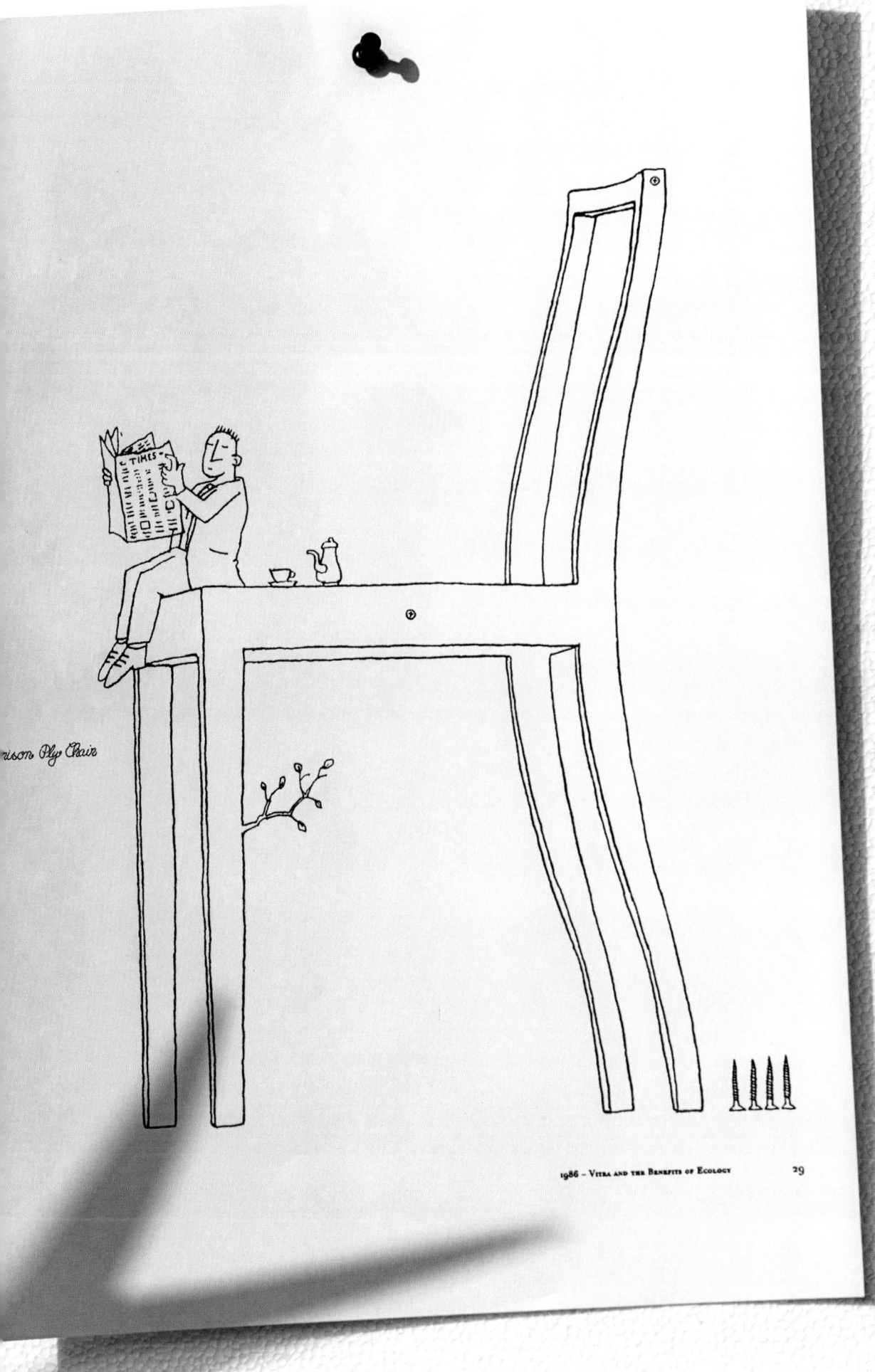

1986 – Vitra and the Benefits of Ecology

GLENN PUSHELBERG & GEORGE YABU

Goofing around in the leather workshop

Shaping the edge of a sofa

The R&D TEAM gets to work on a sofa prototype with Yabu and Pushelberg

MICHELE DE LUCCHI

With his SECRETELLO writing desk (2016)

RODOLFO DORDONI

The vision of elegance demonstrated by the late Rodolfo Dordoni (1954-2023) throughout his career was both unmatched and prolific. He wove the qualities of timelessness, rationality, and refinement into *"total design"* environments – informed by Modernist principles, yet contemporary in their essence – combining skills in architecture and art direction with his expertise in Italian design.

This holistic vision was informed by the many long-standing collaborative relationships Dordoni had across his career with leading Italian designers, brands, and artisans. Born in Milan, he studied architecture at the Polytechnic University of Milan. After graduating in 1979, he dove into the Italian design world, first at Cappellini (until 1989), where he served as art director, then through collaborations with the likes of Foscarini, Flos, and Molteni&C, to name a few. In 2005, he founded Dordoni Architetti with Luca Zaniboni and Alessandro Acerbi to pursue various spatial projects, including exhibition design and architecture and interiors for residential, commercial, and hospitality projects, as well as for yachts and cruise ships.

Dordoni's relationship with Molteni&C started in 2002 and resulted in products that over the years formed families, set moods, and defined directions. Notably, he blended Italian craftsmanship with the poised atmosphere of historic London members' clubs with his Chelsea (2015), Kensington (2018), and Barbican (2018) chairs. The consistency of proportion found in his 606 (2010), 5050 (2016), and 7070 (2022) drawer unit series expanded over time to beds and dining and coffee tables. Dordoni's meticulous attention to detail across scales – from upholstery stitching to spatial atmospheres – was for him a discipline of pride and sincerity, and a distinctive characteristic of the very best of Italian design. –H.T.

JASPER MORRISON

Refreshing simplicity and character are the essence of each object designed by Jasper Morrison (born in London in 1959). Driven by the goal of improving everyday domestic life, he has an extensive portfolio of design work spanning tableware, lighting, electronics, and furniture, and has communicated his astute design perspective through several books, including *The Good Life* (2014) and *A Book of Things* (2015).

 Morrison studied design at Kingston Polytechnic, the Royal College of Art, and HdK Berlin. He established his own studio in London in 1986, making small batches of design pieces, such as the Thinking Man's chair (1985), the Plywood chair (1988) and the 1144 handle (1991), and working on collaborations with design brands such as Alessi, Flos, and Vitra. Style emerges through essentialism in the work of Morrison. It can be seen in his lightweight, solid wood Tea chair (2021) for Molteni&C, which blends comfort and sophistication, as well as in his curation of *"80!Molteni,"* a 2015 exhibition celebrating 80 years of Molteni&C's design collaborations at the Galleria d'Arte Moderna.

 He has further developed his approach through *"Super Normal,"* in collaboration with Japanese designer Naoto Fukasawa, a philosophy on designing everyday tools driven by our unconscious behavior. In 2009, he opened the Jasper Morrison Shop at his London studio, and he has additional studios in Paris and Tokyo. His work is in the collections of design museums around the world, and his contributions to design have been recognized in the form of an Isamu Noguchi Award (2015), the Compasso d'Oro (2020), and a designation as a Commander of the Order of the British Empire (2020). —H.T.

GEORGE YABU AND GLENN PUSHELBERG

George Yabu and Glenn Pushelberg know how to capture the zeitgeist. Synonymous with the most beguiling interiors in the world of luxury hospitality, the Canadian duo is recognized for shepherding the future of hospitality design. Their design studio, Yabu Pushelberg, creates layered atmospheres that spark universal appeal, joy, and delight.

    Partners in work and in life, Yabu and Pushelberg opened their practice in 1980 in Canada. Today, the firm operates from both Toronto and New York, and employs a staff of 110. Through its nearly five-decade-long history, Yabu Pushelberg has evolved from an interior design studio to a comprehensive design practice that includes building design, product design, lighting, styling, textiles, graphics, and strategy. Since their first interiors project for Club Monaco in Toronto in 1984, they have gone on to work on retail spaces for Bergdorf Goodman, Louis Vuitton, and Tiffany & Co.; hotels and resorts for The Four Seasons and Park Hyatt; residences for Aman; and collections for the likes of Glas Italia, Salvatori, and Molteni&C.

    Yabu and Pushelberg's relationship with Molteni&C began a decade before the debut of their first collection together. Understanding the potential of the collaboration, the pair waited until they had a concept that felt right. That design was Surf (2019), a modular sofa system inspired by the views of the Atlantic Ocean seen from the couple's Amagansett, New York, residence. Following Surf, the studio devised Tivalì 2.0 (2022), a modernized iteration of the 2004 design by Dante Bonuccelli, with enclosing aluminum concertina doors.

    Beyond their numerous recognitions across the design world, Yabu and Pushelberg are officers of the Order of Canada (2014) and have founded a scholarship fund and award for innovation at Toronto Metropolitan University, as well as a scholarship at New York's Parsons School of Design. —H.T.

MICHELE DE LUCCHI

Multitudes exist within the Milan-based architect and designer Michele De Lucchi (born in Ferrara, Italy, in 1951). A pivotal player in the Alchimia and Memphis radical Italian design movements, De Lucchi operates today within an experimental, laboratory-esque style, combining disciplines to imagine a human-oriented future.

    De Lucchi studied architecture at Florence University in the 1970s under Superstudio member Adolfo Natalini and cofounded the early radical design group Cavart in 1973. Later, he was a part of the Memphis Group alongside Ettore Sottsass. From 1988 to 2002, De Lucchi served as design director at Olivetti and has collaborated throughout his career with leading brands such as Alessi, Hermès, and Molteni&C, creating the Materic kitchen for Dada in 1999 and the MDL System–in partnership with Angelo Micheli and Giovanni Battista Mercurio–for UniFor from 2003-04. He has also worked on commercial architectural projects and workplace concepts for clients including Deutsche Bank, Poste Italiane, and Telecom Italia. In 1990, De Lucchi founded Produzione Privata, an experimental design lab that conceives and produces experimental objects, free of commission, with the utmost freedom of expression. He is also the founder of AMDL CIRCLE, a multidisciplinary group of thinkers and innovators.

    His approach is grounded in the belief that design has a strong relationship to its user, and that, in designing spaces and objects, one also designs behaviors. Take the Secretello desk (2016) for Molteni&C; a workspace, display cabinet, and intimate treasure-trove, it both reveals and conceals the private ambitions of a writer. De Lucchi was appointed as an Officer of the Italian Republic in 2000 for services to design and architecture, and is a member of the Accademia Nazionale di San Luca in Rome and a professor at Venice University and the Polytechnic University of Milan. In 2022, he was honored with the Compasso d'Oro Career Award. —H.T.

TOBIA SCARPA

Presents an image of the MEO chairs (1981), designed by him and Afra Scarpa

TOBIA SCARPA looks at the MHC.3 MISS, a re-edition of the 1986 Miss chair

Archival images of the MONK chair and the Mou table (both 1973), designed by Afra and Tobia Scarpa

Tobia Scarpa with Carlo Molteni

MICHAEL ANASTASSIADES

Holds an image of his HALF A SQUARE table (2020)

FRANCESCO MEDA

Shows a 3D-printed model of his WOODY chair (2018)

The WOODY chair marked Francesco Meda's debut as a designer for Molteni&C

NAOTO FUKASAWA

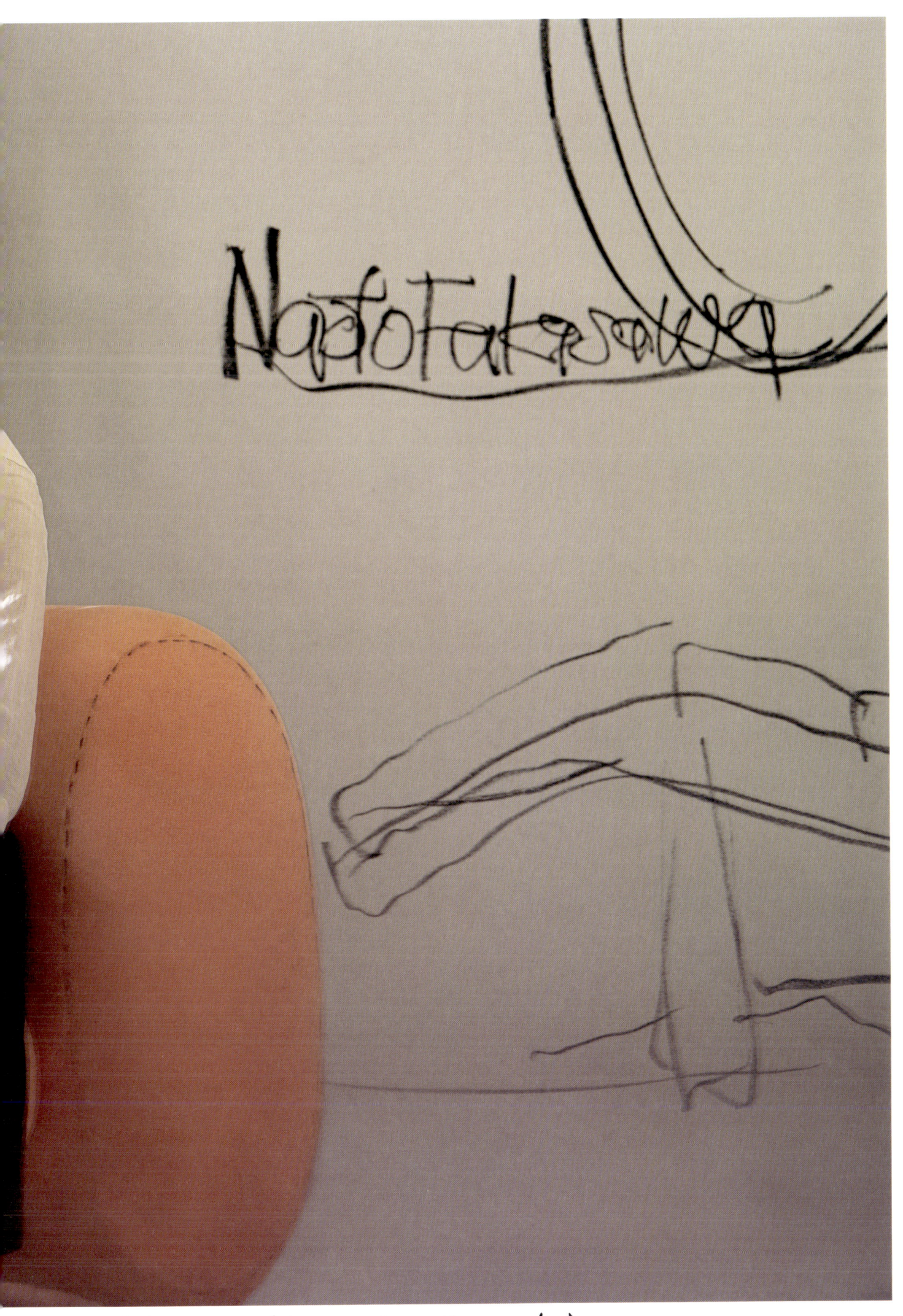

sitting on his CINNAMON armchair

Fukasawa sketches the CINNAMON...

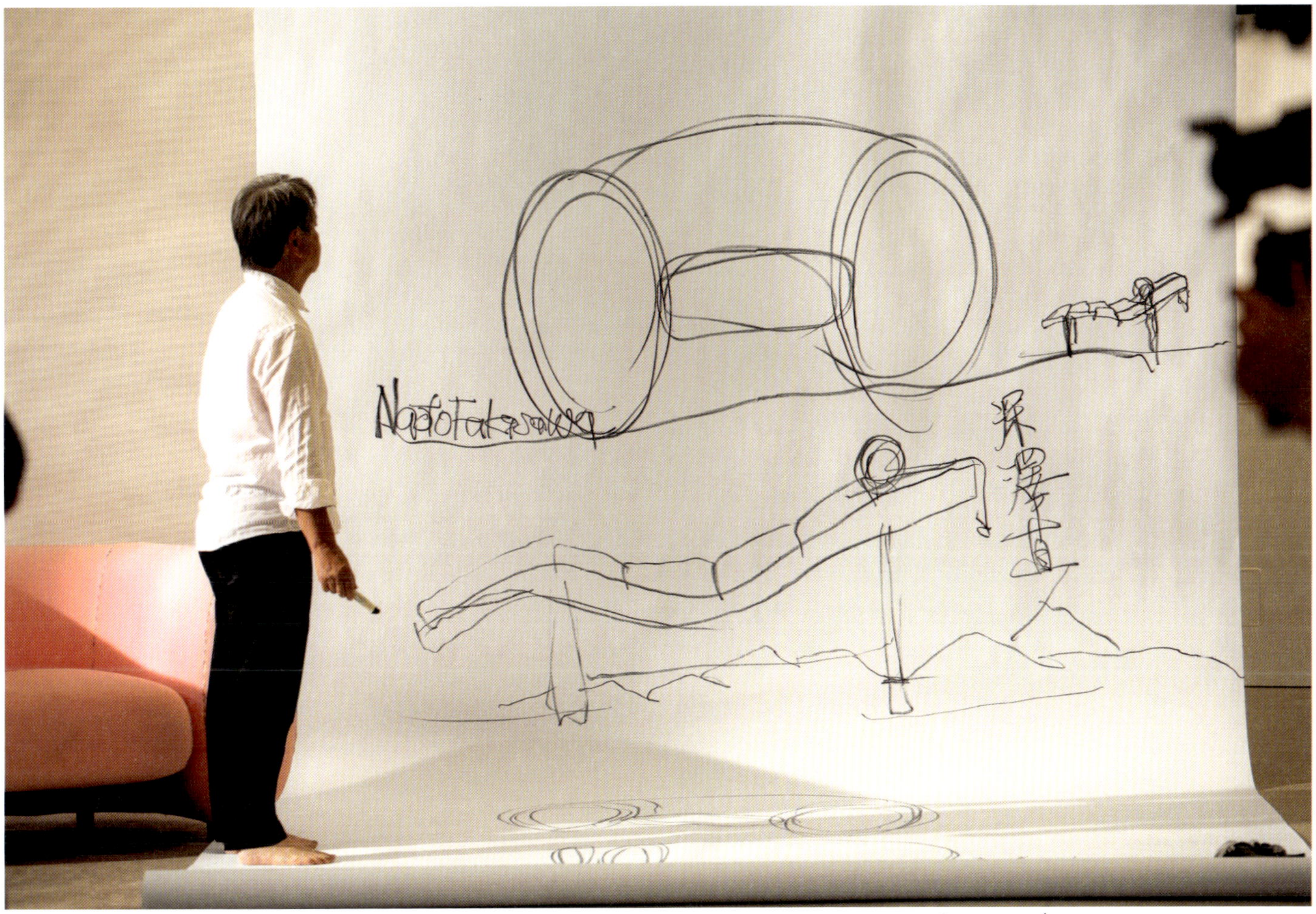

...and the TUSCANY chairs (2023)

RON GILAD

Ron Gilad stands with his TEOREMA drawer units (2017)

Gilad, shown here in front of his 45° table collection (2012), is the curator of the MOLTENI MUSEUM

TOBIA SCARPA

A leading figure in Italian design and architecture, Tobia Scarpa (born in Venice in 1935) blends modernism with technology through designs informed by the history of craftsmanship and heritage. The son of the architect Carlo Scarpa (1906-78) and the husband of the late Afra Scarpa (1937-2011), his longtime partner in both work and life, Tobia has worked across design, interiors, architecture, and restoration throughout his illustrious career.

He studied architecture at Università Iuav di Venezia, where he met Afra, and after graduating, they opened an office together in 1960 in Montebelluna, Italy. While Tobia was working as a glass designer at Venini's Murano workshops, they began designing furniture together, first for Gavina, and later for firms like Flos, San Lorenzo, and Molteni&C. Past interior and architecture projects include the factories and showrooms of Benetton (a collaboration that began in 1964), the showrooms of UniFor, and historical buildings in northern Italy.

Tobia and Afra began their collaboration with Molteni&C in 1972, launching the Monk dining chair (1973), which took on the expressive quality of its materials: tubular steel, Italian walnut, ash, leather and canvas. Further modern designs followed, including the Morna bed (1973), the Mop bookcase (1974), the Mix office chair (1975), the Mastro chair (1980), the Meo chair (1980), the Mita cabinet (1985), the Miss chair (1986), and the Filo chair (1987). In addition to designing furniture, the Scarpas also designed Molteni&C's showrooms in Milan, Rome, and Paris.

Tobia and Afra received the Compasso d'Oro in 1970 (Tobia received it a second time in 2008) and have designs in global collections including New York's Museum of Modern Art, the Philadelphia Museum of Art, London's Victoria & Albert Museum, and the Museum of Contemporary Art Chicago. Between 2001 and 2009, Tobia taught at the University of Venice and the University of Alghero. —H.T.

MICHAEL ANASTASSIADES

Conceptual art, geometry, and function find balance in the designs of the London-based Michael Anastassiades (born in Cyprus in 1967). His minimalist lighting compositions perform as contemporary chandeliers, elevating space with cutting-edge abstractions of nature, mobility, shadow, and light. Treading an exquisite line between art and industry, the essential rigor underlying his design practice can be traced back to his foundational training, first as a civil engineer at London's Imperial College, then as an industrial designer at the Royal College of Art. Early-career experiments combined light with sound and movement with performance. After founding his eponymous studio in the U.K. capital in 1994, both trajectories led to product, furniture, and spatial design, all of which he began to manufacture in-house from 2007.

Collaborations with leading global design brands and manufacturers followed, among them Molteni&C. Anastassiades's approach is captured by his Half a Square table series (2020), the most iconic of which came in Verde Alpi marble and a green aluminum frame. These tables exhibit both his imaginative grasp of composition and his shrewd ability to harness the inherent qualities of materials.

Both skills can be traced back to his childhood in Cyprus in the 1970s and '80s, when, isolated during postwar recovery, he built his own worlds in the coastal island landscape. Today, his work can be found in the collections of New York's Museum of Modern Art, the Art Institute Chicago, the Victoria & Albert Museum in London, and the MAK Vienna. He is a Royal Designer for Industry at the Royal Society of Art and has lectured at institutions such as the Camberwell College of Arts, Central Saint Martins, the University of the Arts London, and the École cantonale d'art de Lausanne. In 2023, he was awarded an OBE by King Charles III. —H.T.

FRANCESCO MEDA

The intrinsic qualities of natural materials are the starting point for designer Francesco Meda (born in Milan in 1984). By leaning into the strengths and limitations of marble, brass, or glass, he draws out their natural beauty with artisanal and digital methods, shaping them into functional, modern objects.

Meda studied industrial design at Milan's Istituto Europeo di Design, and after graduating in 2006, worked in London for two years in the studios of Sebastian Bergne and Ross Lovegrove, learning how to bring ideas to life, from crafting a vision to managing a production cycle. He returned to Milan in 2008 to work in the studio of his father, Alberta Meda, an industrial designer and mechanical engineer who, across his long career, collaborated notably with Alessi, Olivetti, Vitra, and others. As an assistant, Meda developed his material-led approach, and though the transmission of generational knowledge continued, he began to experiment on his own with digital technology, including 3D printers, scanners, and parametric software.

He established his own studio in 2012, producing limited-edition objects and collections of jewelry, lamps, and furniture, and collaborating with leading design brands including Molteni&C. His ability to harmonize materials can be seen in his collaborations with Molteni&C: Woody (2018), a gently ergonomic chair with a seamless solid wood frame, and Sistema XY (2020), a modular kitchen system inspired by Cartesian coordinates that encourages smooth organization.

Meda won the Compasso d'Oro 2016, and his work is featured in the Triennale Design Museum collection. He has collaborated with the fashion brands Tod's and Ferragamo, as well as with the Milan galleries Nilufar and Rossana Orlandi.   —H.T.

NAOTO FUKASAWA

According to Naoto Fukasawa (born in Yamanashi Prefecture, Japan, in 1956), truly intuitive design goes unnoticed. It's a quality achieved through simplicity of form, absolute efficiency of purpose, and an innate understanding of human nature – values he advocates for explicitly through writing, teaching, and curating, and implicitly via his works of design, from electronics and watches and pens to furniture and his L-shaped house and studio in Tokyo. Driven by the humanist values of longevity and accessibility, Fukasawa has frequently worked in the field of technology. He earned his degree in product design at Tama Art University in 1980 and began his career at Seiko Epson in microelectronics before moving to San Francisco in 1989 to work at IDEO in Silicon Valley. In 1996, he returned to Japan with a desire to reconnect with the country's aesthetics, inspired in particular by the work of the Japanese-American artist and designer Isamu Noguchi (1904-88).

In 2003, he founded Naoto Fukasawa Design, undertaking collaborations with leading design brands such as Alessi and Artemide, resulting in discerning products – some of which now feature in the collections of New York's Museum of Modern Art, London's Victoria & Albert Museum, and the Designmuseum Denmark. In 2006, along with Jasper Morrison, he established *"Super Normal,"* a grounding philosophy for designing everyday tools informed by unconscious behavior. For Molteni&C, he has designed the Cinnamon armchair (2023), a bulbous seat on a U-shaped base that molds to the body, and the Tuscany chaise lounge (2023), which combines Japanese and Italian aesthetics with a fluid silhouette and contoured structure of solid wood and brass.

Fukasawa is the director of the Japan Folk Crafts Museum and 21_21 Design Sight, both in Tokyo; a vice president at Tama Art University; and a judge for the Loewe Craft Prize. He was chairman for the Japan Good Design award from 2010 to 2014 and received the Isamu Noguchi Award in 2018. — H.T.

RON GILAD

Artist Ron Gilad (born in Tel Aviv in 1972) uses design to subvert perceptions about space. Hovering between the abstract and the functional, his hybrid creations, from sculpture to furniture to interiors, fuse humor, minimalism, and poetry, whether as limited-edition art or as mass-production designs. His particular alchemy is a product of his cross-disciplinary education and his playful personality.

Gilad studied architecture in high school, industrial design at the Bezalel Academy of Art and Design in Jerusalem, and right after graduating, started lecturing on 3D and conceptual design at Shenkar Design & Engineering Academy in Ramat-Gan, Israel, with a focus on art jewelry. Moving to New York in 2001, he became a professor at the Pratt Institute in Brooklyn and set up his design firm, now based between Tel Aviv and Milan. Under his studio, Gilad has created diverse, maverick objects—such as stools, bowls, and lighting fixtures—several of which are now in the permanent collections of New York's Museum of Modern Art and Metropolitan Museum of Art, and of the Tel Aviv Museum of Art. In 2008, he began collaborating with Italian design companies, to date working with the likes of Flos and Molteni&C. Since 2016, Gilad has also been a visiting professor at both the Polytechnic University of Milan and the Nuova Accademia di Belle Arti in Milan.

His fertile collaborations with Molteni&C, which marry furniture and spatial design, have won numerous awards. The Teorema chest of drawers (2017) is a piece of *"micro architecture"* in which stacked volumes and drawers in wood can be rotated by 20 degrees. The Grado° collection (2012) of consoles, tables, mirrors, and shelves questions perceptions of walls and space, using pure geometric angles and the invisibility of glass. Gilad is also the designer of the Molteni Museum (see pages 303-317), located at Molteni&C's headquarters, where he curates whimsical and provocative installations. —H.T.

MOLTENI MONDO

SCENE | SLATE | TAKE

THE ARCHIVE

ROLL

DIR / DOP JEFF BURTON

DATE 1934—2024

"OUR WILLINGNE
DESIGNERS
POSSIBLE SOLU
ONE OF OUR S
BUT ALSO TH
PART OF

ss TO WORK WITH
 FIND EVERY
ON IS NOT ONLY
TRONG POINTS
MOST EXCITING
OUR WORK."

CARLO MOLTENI

THE ARCHIVE

By Emma Leigh Macdonald

It may be difficult to imagine today, when objects like tables and chairs can seemingly be made in any shape imaginable, but part of Molteni&C's predominant influence in the world of design has been its revolutionary advances in fabrication technology, allowing for the creation of items that would have been impossible to produce with the limited materials and techniques available before the Second World War. From Afra and Tobia Scarpa's clean-lined Monk chair (1973) and striking Mop bookcase (1974) to today's walk-in wardrobes and bookcase systems that redefine what constitutes a partition in a space, the company has long balanced innovation with tradition.

One way this dual reverence for both the past and the future is evident is through Molteni&C's careful attention to its archive. The history of this history, you could say, was born in 2013, when the company first started planning its eightieth-anniversary celebrations, exhibition, and catalog. As the family and the larger Molteni&C team began to organize their collection of materials for the occasion—including more than 1,600 drawings, 1,200 catalogs, hundreds of documents, annotated slides, prototypes and original objects, videos, ads, and personal materials—the original Molteni Museum came to life.

Today, the Museum serves as a beautiful home for these important materials. While it was first designed by Jasper Morrison and located atop the company's Giussano showroom, the Museum was relocated in 2021 to Ron Gilad's Glass Cube in the gardens of the Molteni Compound (pages 20-63). Inside this transparent space, visitors, students, architects, and clients can view the company's trajectory—not only its most standout designs, many of which are still in production today, but also how the family and the designers with whom they have collaborated over the years arrived at these category-defining creations. A prototype that has been preserved may never have become mass-produced, for example, but perhaps one detail from that original design can be traced to another piece that made it into living rooms all over the world.

     Angelo Molteni, the company's visionary founder, and his wife, Giuseppina, were two of the original creators of the Salone del Mobile design and furniture fair in Milan – its first iteration held in 1961 – which further built upon Molteni&C's legacy of collaboration and connection. Through projects with Luca Meda in the late 1960s; Aldo Rossi in the 1980s and '90s; Jean Nouvel in the '90s; Foster + Partners, beginning with the Arc table in 2009; and, more recently, Michael Anastassiades, to name a few, Molteni&C became a design-world paradigm composed of a who's-who of *"your favorite designer's favorite designer."* Along these lines, Molteni&C's reissue of furniture designed by Gio Ponti between 1935 and the 1970s – including the Round D.154.5 chair, initially designed in the early 1950s and revived by Molteni&C in 2021 – is another instance of design preservation, resulting in key work being brought back into the present. In the archive, certain more intimate moments behind these collections are preserved, as well: in a well-known video interview with Rossi from 1991, the architect appears relaxed and makes light-hearted commentary between answering questions about his new work.

     Molteni&C's historical importance can be attributed, in part, to its prioritization of craftsmanship over style – the former which, ironically enough, never goes out of style. The company has stayed relevant due to an unwavering loyalty to its design heritage and know-how, but also because it's one to continually push boundaries. The Pavilion (pages 36-39), opened in 2022, no doubt exemplifies the continued excellence of this ahead-of-the-curve company. The Museum's position inside a glass house puts Molteni&C's historic works squarely in view amidst the bustling campus, creating a perfect visual metaphor for the company as it enters its next decade and beyond.

Molteni & C circa 1960s furniture pieces for the living room

Wardrobe from the FIORENZA series (1968), designed by Tito Agnoli

Classic Molteni & C drawer unit from the 1960s before the company started its design production

Aldo Rossi and Luca Meda's TEATRO (1982), Meda's RISIEDO (1988), and Meda and Rossi's PIROSCAFO (1991)

Luca Meda

In the '60s, Molteni&C moved from traditional furniture produc. (as seen in the background).

... to design, shown here in the foreground Tito Agnoli's FIORENZA collection (1969)

In the foreground: Molteni&C campaign on FIORENZA (original photo by Aldo Ballo)

Luca Meda and Hans von Klier's IRIDE modular storage system (1968)

Luca Meda and Hans von Klier's IRIDE storage system (left), along with Meda's 7VOLTE7 system (1988)

MANIFESTO, as installed at Salone del Mobile in 1990, showcasing Molteni &C and

UniFor furniture by Aldo Rossi, Original photo by Mario Carrieri

Left: Bookshelf prototype designed by Yasuhiko Itoh. Received third prize at the Terzo Concorso Internazionale del mobile di Cantù

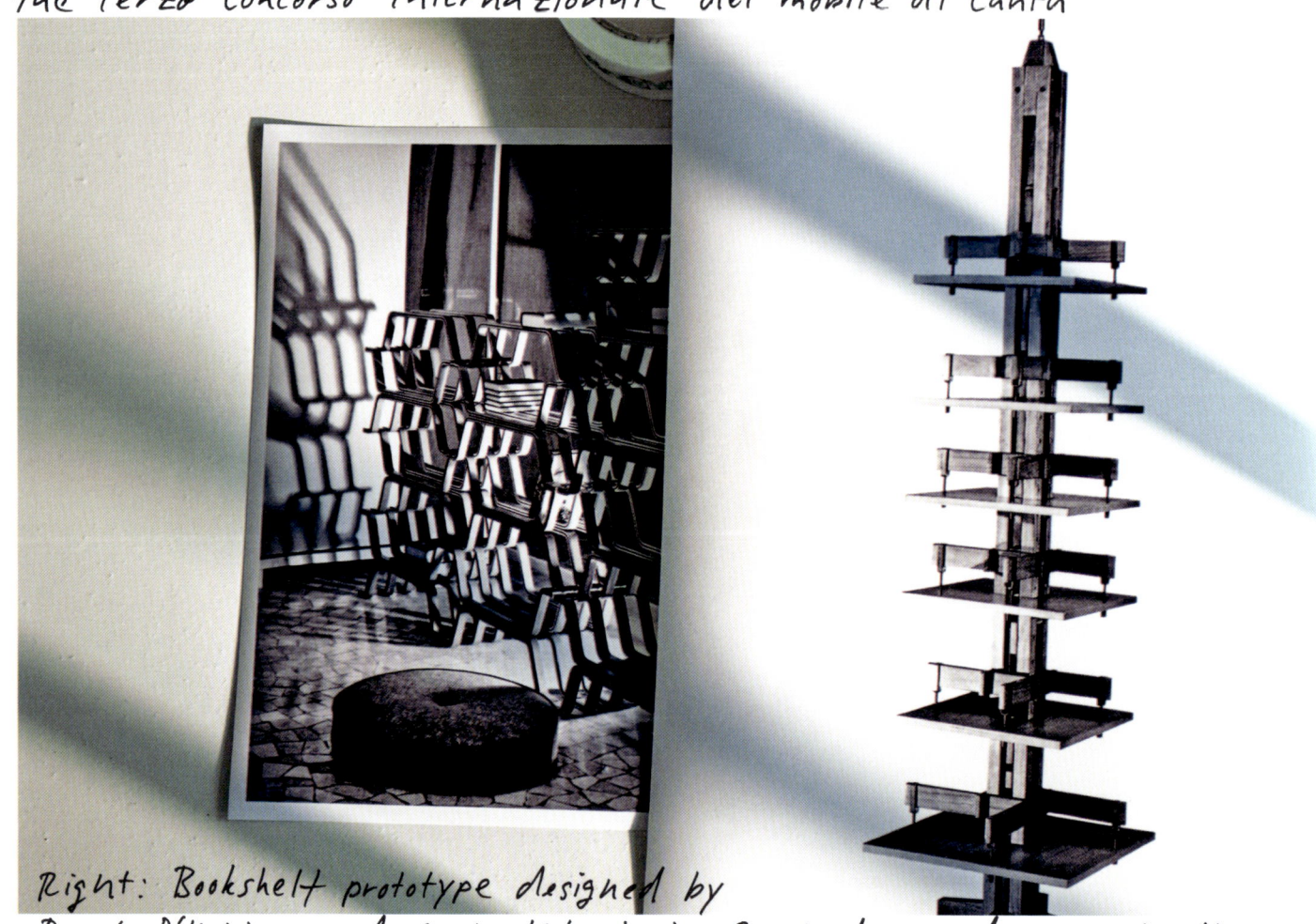

Right: Bookshelf prototype designed by Donato D'Urbino and Carlo Volonterio. Received second prize in the student category at the Terzo Concorso Internazionale di Cantù in 1959

A drawing of Luca Meda and Franco Giacometti's RASTER series from the 1980s

Right: A sketch by Luca Meda for the concept of the 1993 presentation of PIROSCAFO at Salone del Mobile

Hannes Wettstein's ALFA chair (2001)

Sketches of the VIVETTE armchair (1989) from Luca Meda

Salone del Mobile poster (1981)

Sketches and pictures of Molteni&C collaborations with Aldo Rossi (left) and Luca Meda

Drawing and prototype for Rossi and Meda's PLATEA chair (1990) designed for Teatro Carlo Felice in Genoa, Italy, by UniFor and Molteni&C

Poster to promote the PRIMAFILA sofa (1990-91) designed by Luca Meda for the foyer of Teatro Carlo Felice in Genoa, Italy

Ò
disegnato questo
divano per il foyer del teatro
Carlo Felice di Genova: l'ho disegnato
PENSANDO A IGNAZIO GARDELLA E AD ALDO ROSSI,
cioè a coloro che sono per me maestro e amico.
Volevo, sia pur attraverso un mobile, interpretare
IL SENSO LONTANO DEL LORO LAVORO, RICERCANDO
un rapporto, che da sempre è comune ad entrambi,
con la tradizione e la ragione. Sono sempre stato
convinto che non vi sia razionalità possibile per

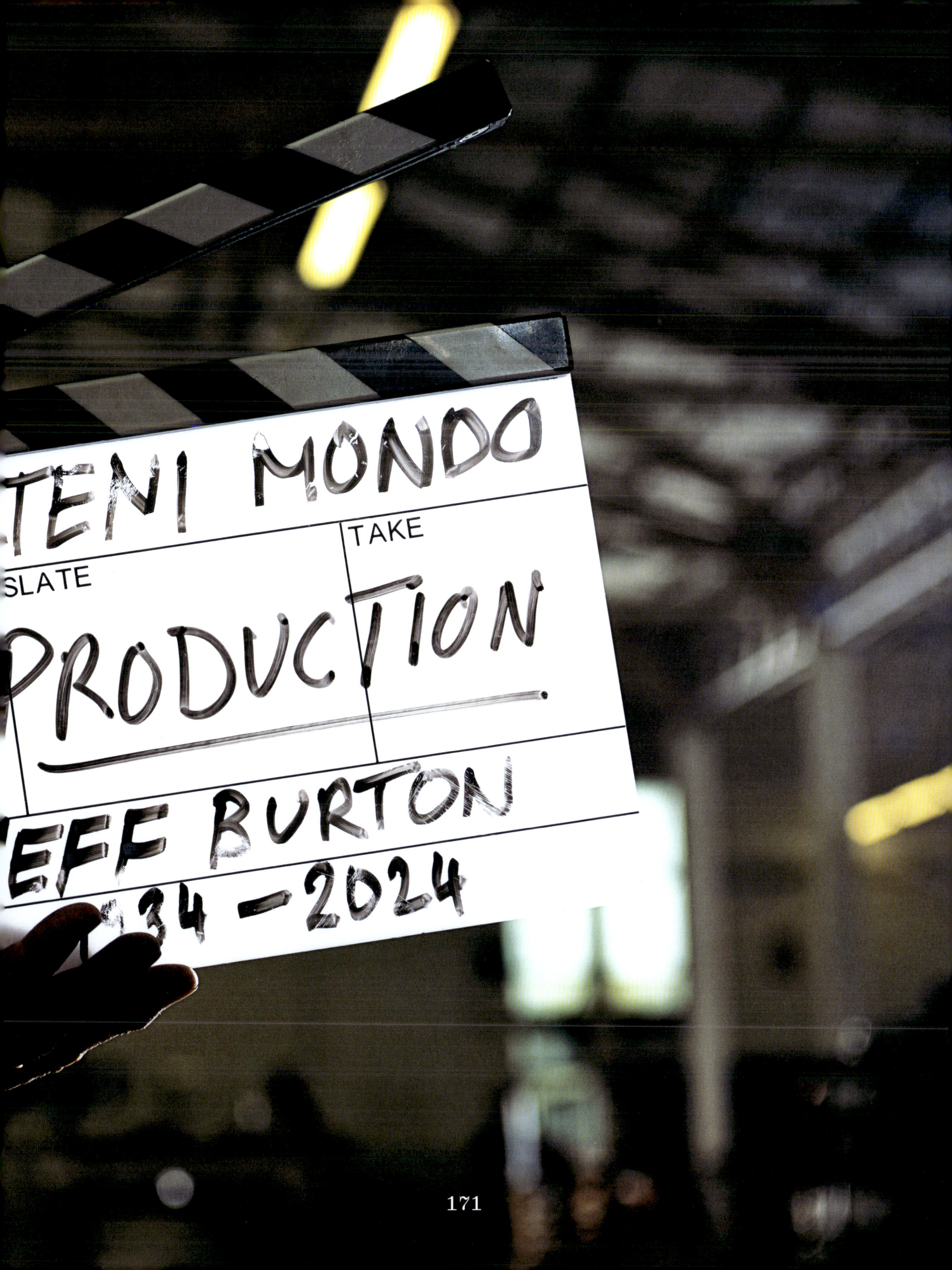
TENI MONDO
TAKE
SLATE
PRODUCTION
EFF BURTON
34 – 2024

CARLO MOLTENI — Rides his bike in the plate and veneer production area of the factory

Checking veneer sheets for continuity of color and veins

Views from inside the veneer production area of the factory

Sewing together a veneered sheet by hand

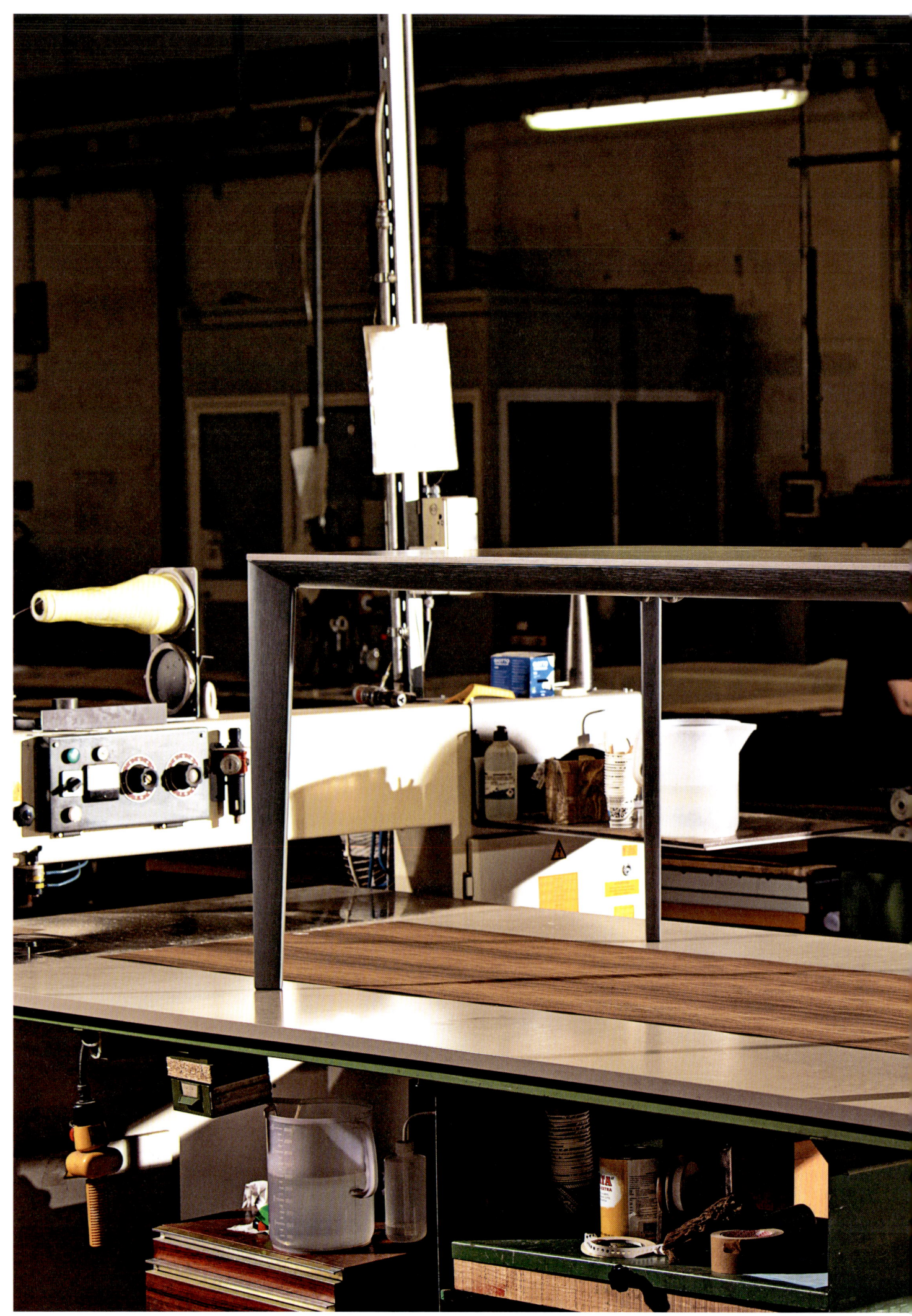

Sewing together veneered sheets

which maintains consistency in color and wood grain in the final product

The coating area of the factory

Robotic Machines paint the wood

A factory worker hand-varnishes solid wood surfaces

Quality control in the painting area of the factory

THE FACTORY

By Francesca Picchi

If the entrepreneurs who gave birth to the model of the Italian design factory share one characteristic trait, it is that of having developed a wholly original approach to production as a job to be respected and revered, something that stems from a far-away time and a tradition to be handed down and updated from one generation to the next – almost as if it were a common good to be taken care of. Perhaps more than anything else, these central figures in the creation of Italian design factories are the ones who have kept the tiller straight in the midst of global transformations and conglomeratization in the twenty-first century, and who have not let industry – *production* – become a neutral tool, ensuring instead that it remains the focus of the creative process and stays available to creative minds for exploring its language and codes.

This phenomenon, so typically Italian, has been made possible by several factors – one of the central ones being the special nature of the industrial fabric, traditionally made up of family-run small and medium-size enterprises – but perhaps the main reason lies in the fact that the companies are grafted in a centuries-old craft tradition. The mentalities of these deep roots have been preserved, despite innovations in technology and production processes, and the pressures to do business on a global scale. This is the same mindset that sets itself the goal of the *"good job well done"* that the London School of Economics professor Richard Sennett identified as *"the craftsman's primordial mark of identity,"* so much so that he made it a category of study and dedicated an entire chapter in one of his best-known books, *The Craftsman* (2008), to *"the obsession with quality."*

In the mid-2000s, together with the designer and curator Paola Antonelli and Kartell founder Giulio Castelli, I set out to recount this chapter so intimately connected to the history of Italian design in the publication *La fabbrica del design* (2007). We tasked ourselves with visiting the entrepreneurs directly at their places of work in order to better understand the many small and large factories scattered around Brianza, famous in Italy for furniture production, and the numerous manufacturing districts surrounding Italy's industrial cities. This brought us to Giussano, the small town near Milan where Molteni&C has been based for nearly a hundred years, a furniture factory that not only had a reputation for being forward-looking and investing in technology, but was known most of all for its control of the entire production cycle, from tree trunk to finished article. It was evidently the only way to guarantee that obsession with quality so essentially artisanal the kind that Sennett writes of in *The Craftsman*–albeit within an industrial context. It was also indicative of the organization of the entire district, since Molteni&C synergistically sold some of its semi-finished products to other local companies. Despite sounding like a mere detail, this gave a glimpse of what appeared to be a vital ecosystem, built on a widespread and measured knowledge of resource conservation. In fact, if there is one trait that distinguishes the character of the people of Brianza, it's their dedication to thrift, one might say, an instinctive intolerance of waste so typical of Lombardy, the unconscious descendant of the motto adopted by the noblest families: *humilitas*.

    As we approached the factory, it became increasingly clear that the two tall towers that stood out from afar because of their bright red bricks were part of Molteni&C. We were told that they were used to store wood-processing waste–shavings and dust–and channel it to the thermal power plant to generate energy. It was striking that the most recognizable part of the architecture openly manifested what is now called *"sustainability."* It was something already evident in the founding act of the factory. (Today, Molteni&C implements every possible effort to meet the ESG goals of the European Commission's 2015 plan to support sustainable growth, including through packaging and by following Forest Stewardship Certificate wood-sourcing guidelines.) As much as the thermal power station, updated to the latest energy standards, and the photovoltaic system installed on the roof pursue the goal of energy autonomy, the two towers continue to be a sign of recognition and mark the identity of the factory, almost as if impersonating the postulate of the eighteenth-century French scientist Antoine Lavoisier: *"Nothing is lost. Nothing is created. Everything is transformed."*

PRODUCTION

By Francesca Picchi

Perhaps not everyone is aware of Brianza's role in the modernization of Italy, but the city's confluence of architects and designers, committed to imagining new living scenarios, and the area's carpenters, willing to follow new directions in style and production, has contributed enormously to the reputation of design in Italy.

Asked whether this phenomenon within the district has changed, Carlo Molteni – the oldest brother representing the second generation of entrepreneurs in the Molteni family – replied this way: *"Yes, Brianza has changed; however, some things have remained the same to this day. Let me give you an example: The foundry that once molded brass feet is still there. Today it makes particularly sophisticated hinges for us, devices with more advanced technologies, but the supplier is the same—as are many of the people I have known since I was a boy and first set foot in the company. There has been an evolution, but in a certain sense we have grown together. Our main suppliers are, in fact, concentrated within a radius of about thirty kilometers [eighteen miles] around our companies, which in turn are all within a few kilometers of one another. It's not just a matter of logistics; it's a choice we made to be immersed in a culture in which we immediately understand one another, or rather have a collective propensity that I would call a taste for craftsmanship. After all, it is the same reason that prompted us to locate our lab, where we work on prototyping and engineering new projects, at the center of the factory. It is the heart of our company, and not only in a metaphorical sense."*

Carlo continued, *"The best-known example is the story of the Less table, developed by Jean Nouvel for the Fondation Cartier in Paris. Nouvel came to the company with a paper envelope in his hand, saying that he was looking for furniture with the same visual permeability of the architecture. He said he wanted a slender construction reminiscent of the essential clarity and subtlety of a folded sheet of paper. After several attempts, we came up with a table of great visual simplicity, one which is still very successful today. Later, we asked him to think about a bookcase, and he said he was interested in designing a lightweight bookcase that did not rest on a pedestal. I was taken aback. I remember him saying, 'I just came from Toulouse, where they manage to make planes fly that weigh tons. Surely we can manage to hold a few books without a base on the ground?' It became a sort of challenge. We put our thinking hats on and worked out a solution. We found it in a small metal element: a perfectly resolved detail on which the entire construction is based. Today, this bookcase is in our catalog under the name Graduate."*

In many respects, the factory seems to function as a kind of tailor's shop for made-to-measure projects. Carlo agrees: "I am convinced that it *is* important to hang onto this willingness to spend time on special projects. It is more significant than you can imagine. In recent years, we have been involved with Herzog & de Meuron on the project for the National Library of Israel–a project on which, as a Group, we have worked with great enthusiasm, putting all of our efforts into the fine-tuning of the design stage to then entrust production to our subsidiary UniFor. Together we have developed a vast range of customized furniture where the two scales–architecture and furniture–merge. This is the case with the wooden bookcase that surrounds the large empty space in the center of the library. It is an element on an architectural scale tasked with defining the space, but also with preserving the books that are the heart of the whole project. The entire design and production phase was an incredible journey. This challenge was complex, one of those opportunities that drives a company to grow. Jacques Herzog and Pierre de Meuron designed the library right down to the smallest detail. On assessing the project for the library reading room–where the most ancient books are stored– they realized that scholars would spend long hours studying here. The specificity of this location prompted them to design the armchairs almost as if they were mini architectures: they wanted to convey the feeling of being in a room, rather than sitting on a simple object. After painstaking refinement and numerous prototypes, they have now become part of our catalog. I must say that our willingness to collaborate with designers to find every possible solution is not only one of our strong points but also the most exciting part of our work."

A panel saw machine, used to cut wood panels

Edgebands used to finish up the sides of wood panels

The edgebanding machine

Previous spread: The aluminum area of the factory

Finishing a glass door

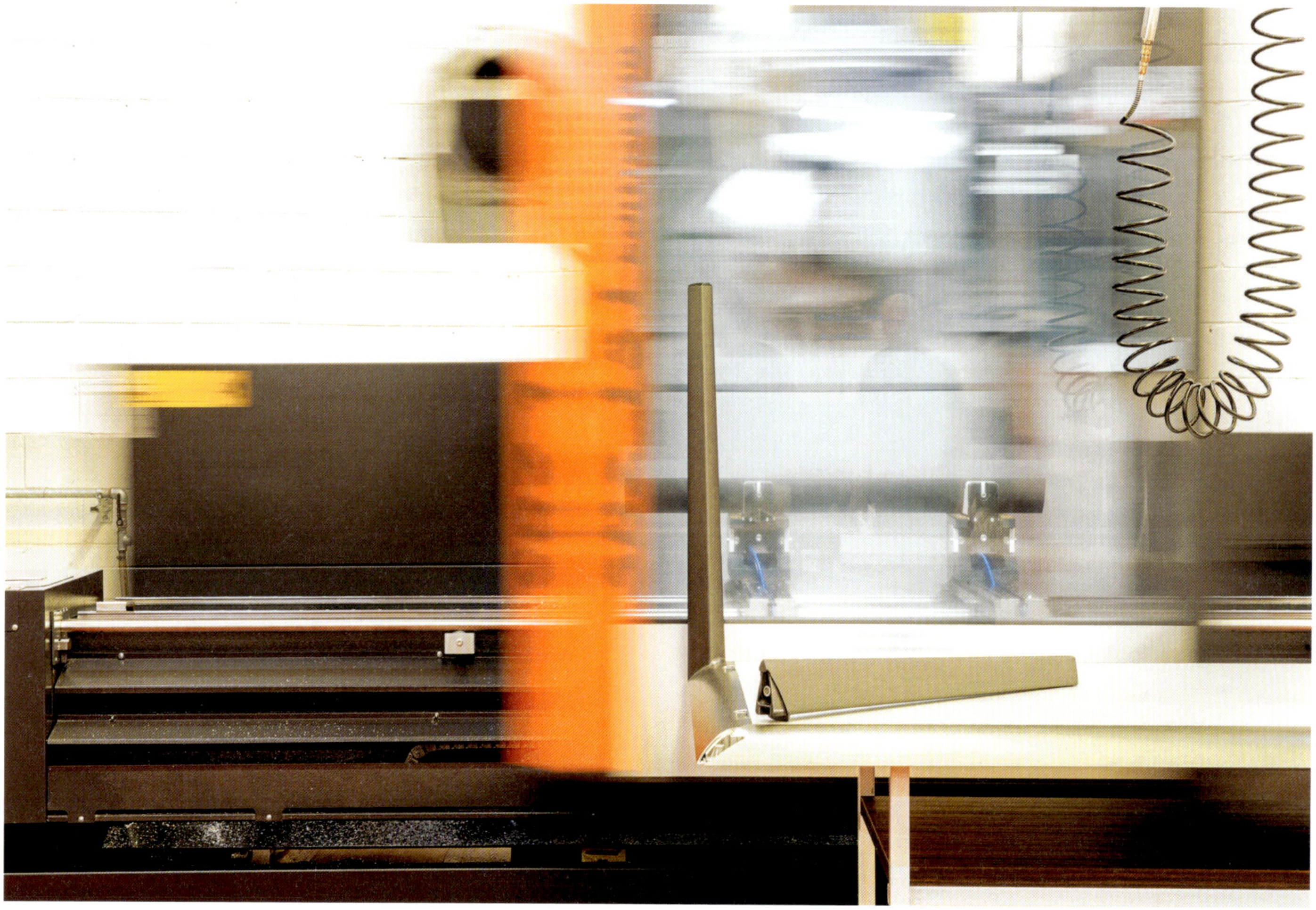

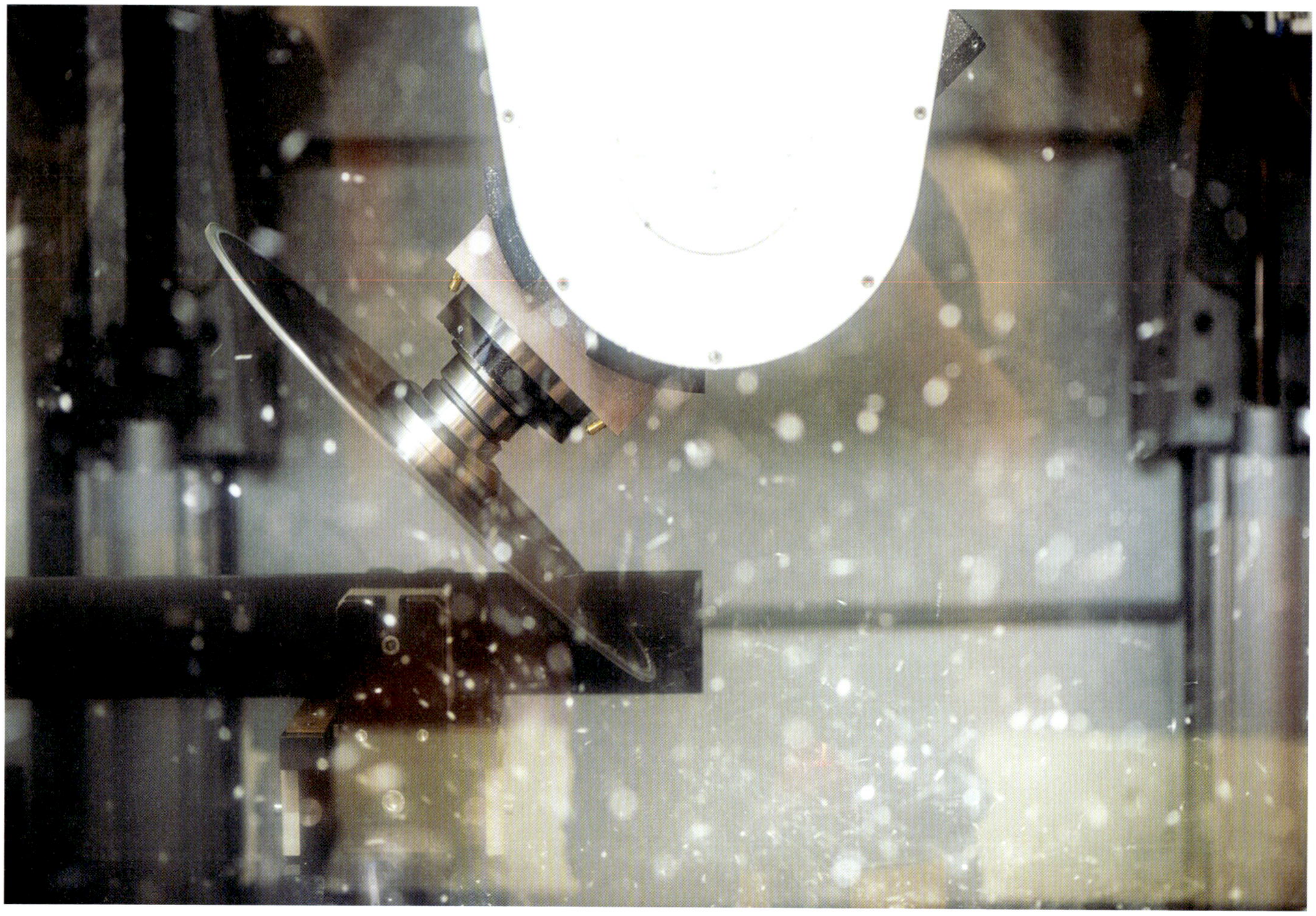

Details of the machine used to cut aluminum

Adding legs to a FILIGREE table (2013)

The area of the factory dedicated to the cutting of leather

A selection of leathers for Molteni&C's upholstery production

The company offers seven types of leather in 110 colors

Molteni & C's carefully chosen PANDORA leather, tanned with chrome and colored with aniline dye

The process of cutting leather for production...

...using laser technologies

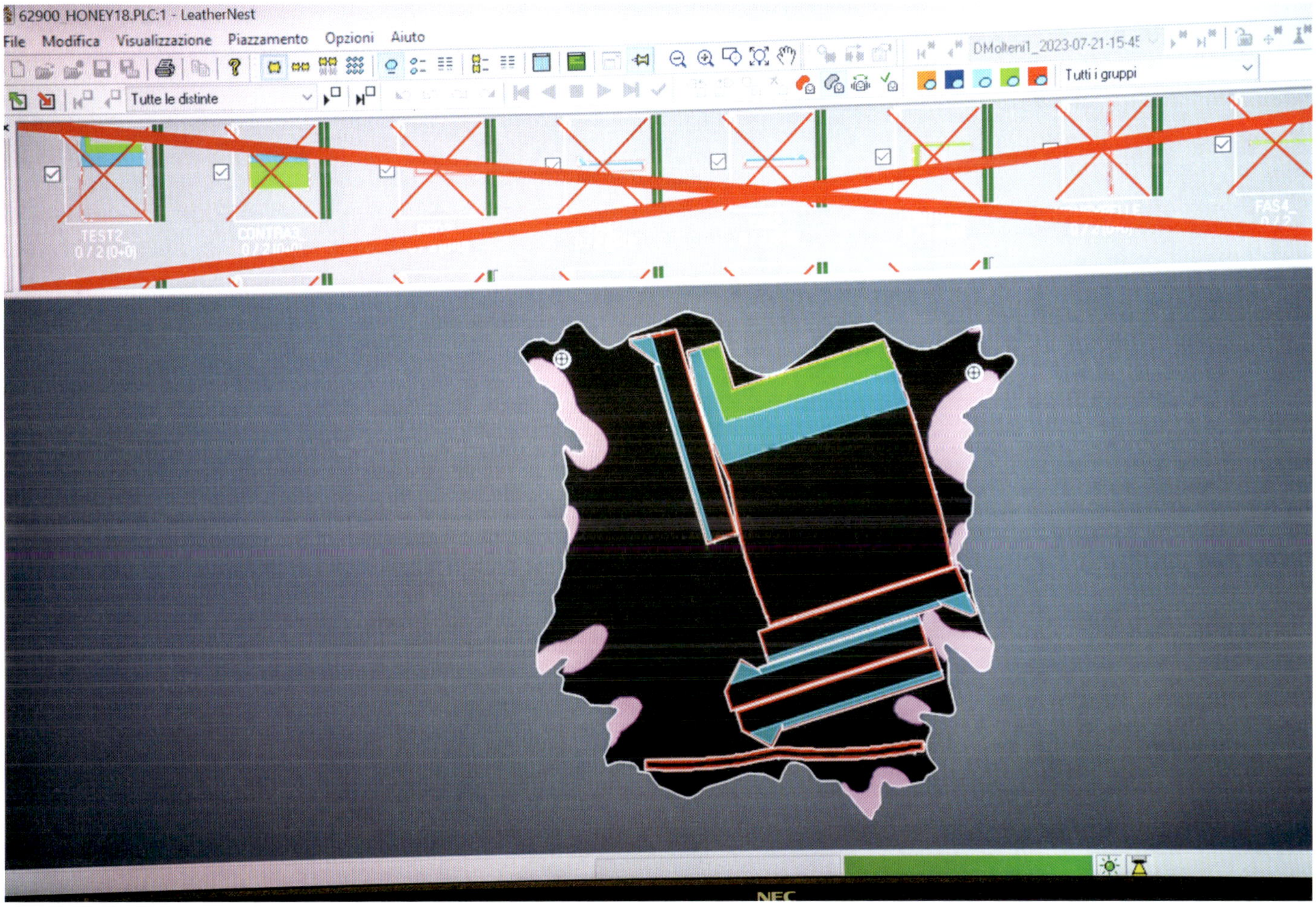

...allows for maximum quality control in the final product and to minimize waste

Positioning a cutting template on leather

The factory's robotic-storage system

The cutting of leather...

...and finishing (or "fleshing") process

Finishing the seams

"IF THERE IS O
DISTINGUISHES
OF THE PEOPL
IT'S THEIR DED
ONE MIGHT
INSTINCTIVE
OF W

NE TRAIT THAT
 THE CHARACTER
 OF BRIANZA
CATION TO THRIFT.
EVEN SAY, AN
INTOLERANCE
ASTE."

FRANCESCA PICCHI

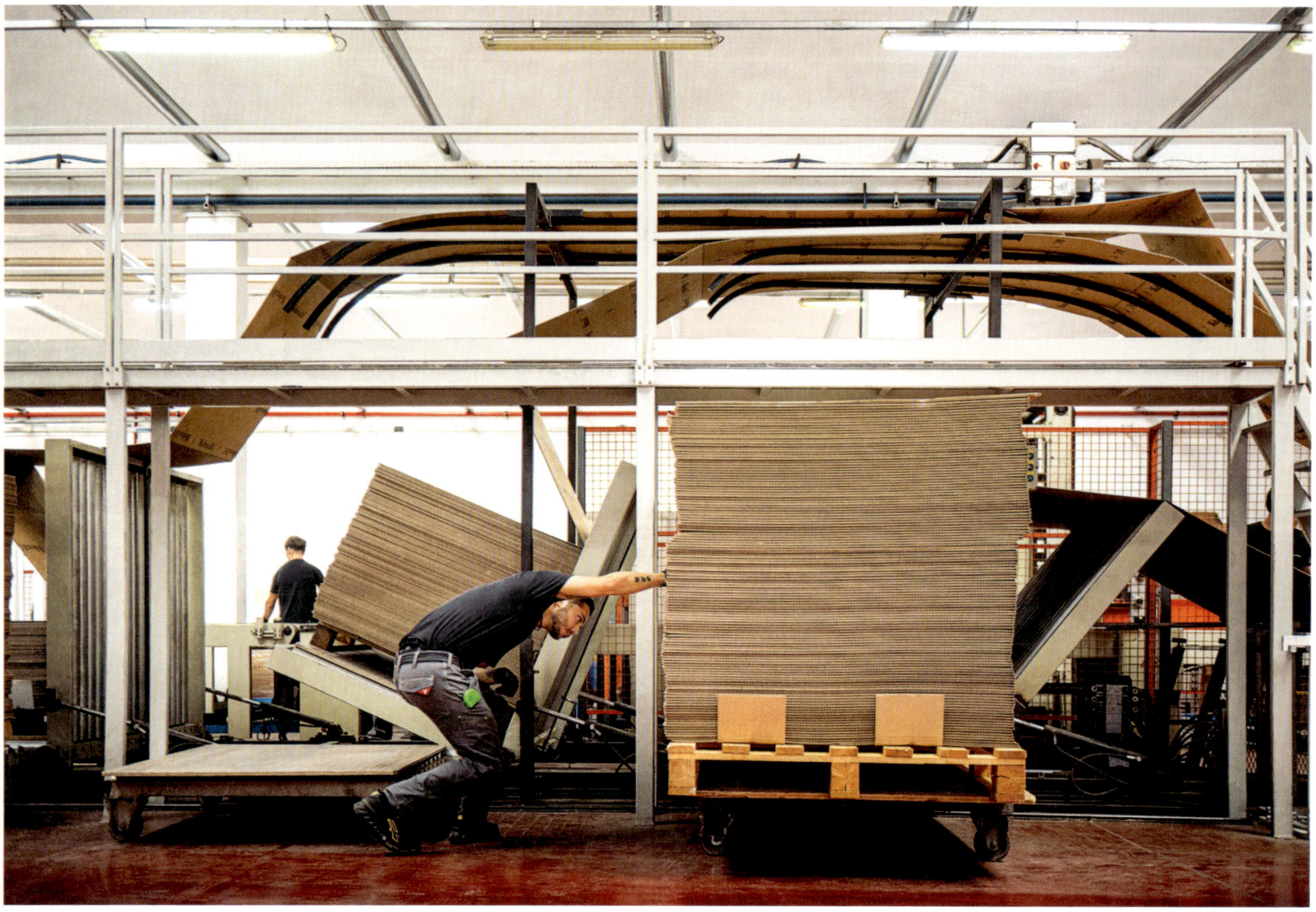

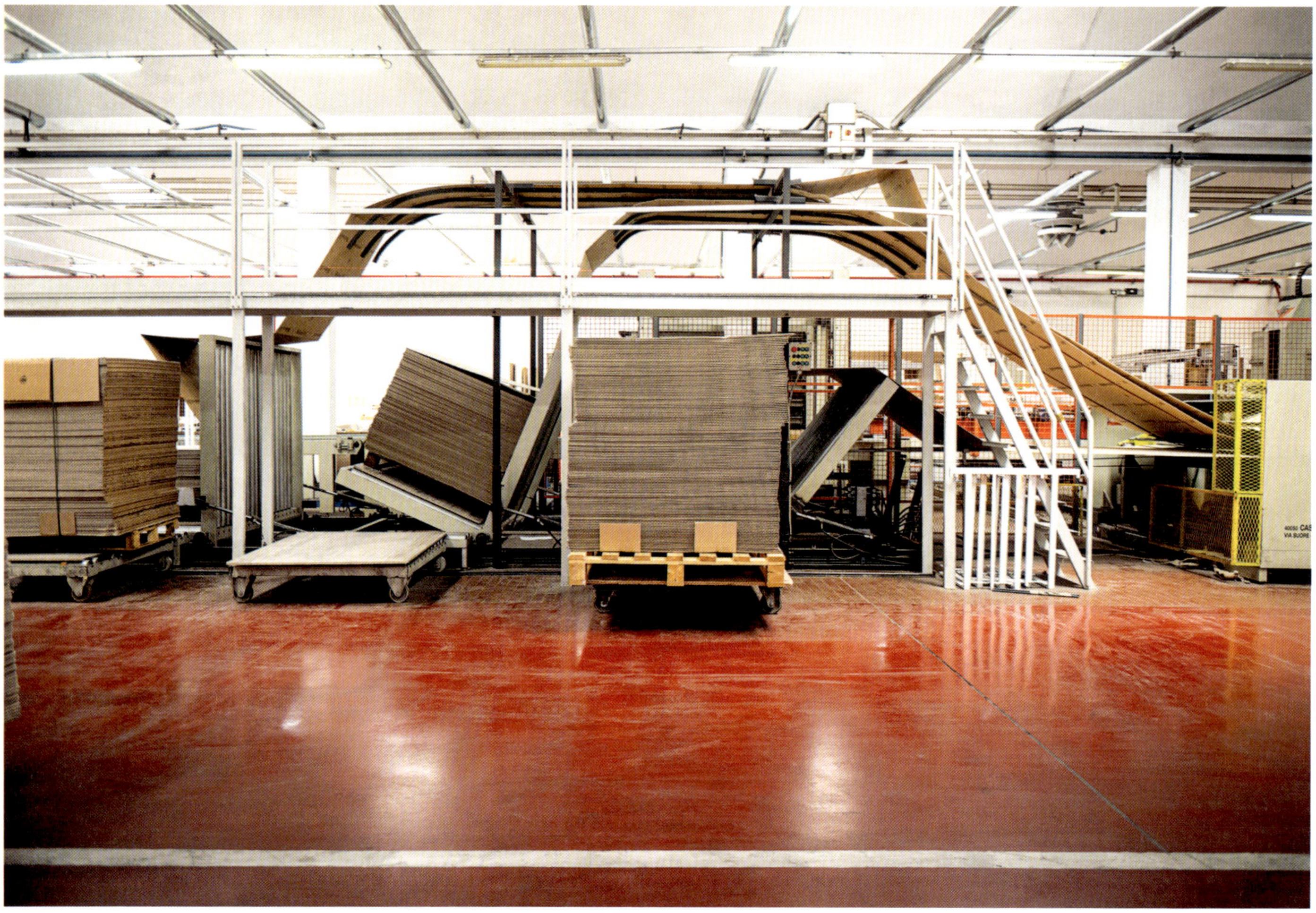

The factory's packaging area, with tailor-made boxes for each piece

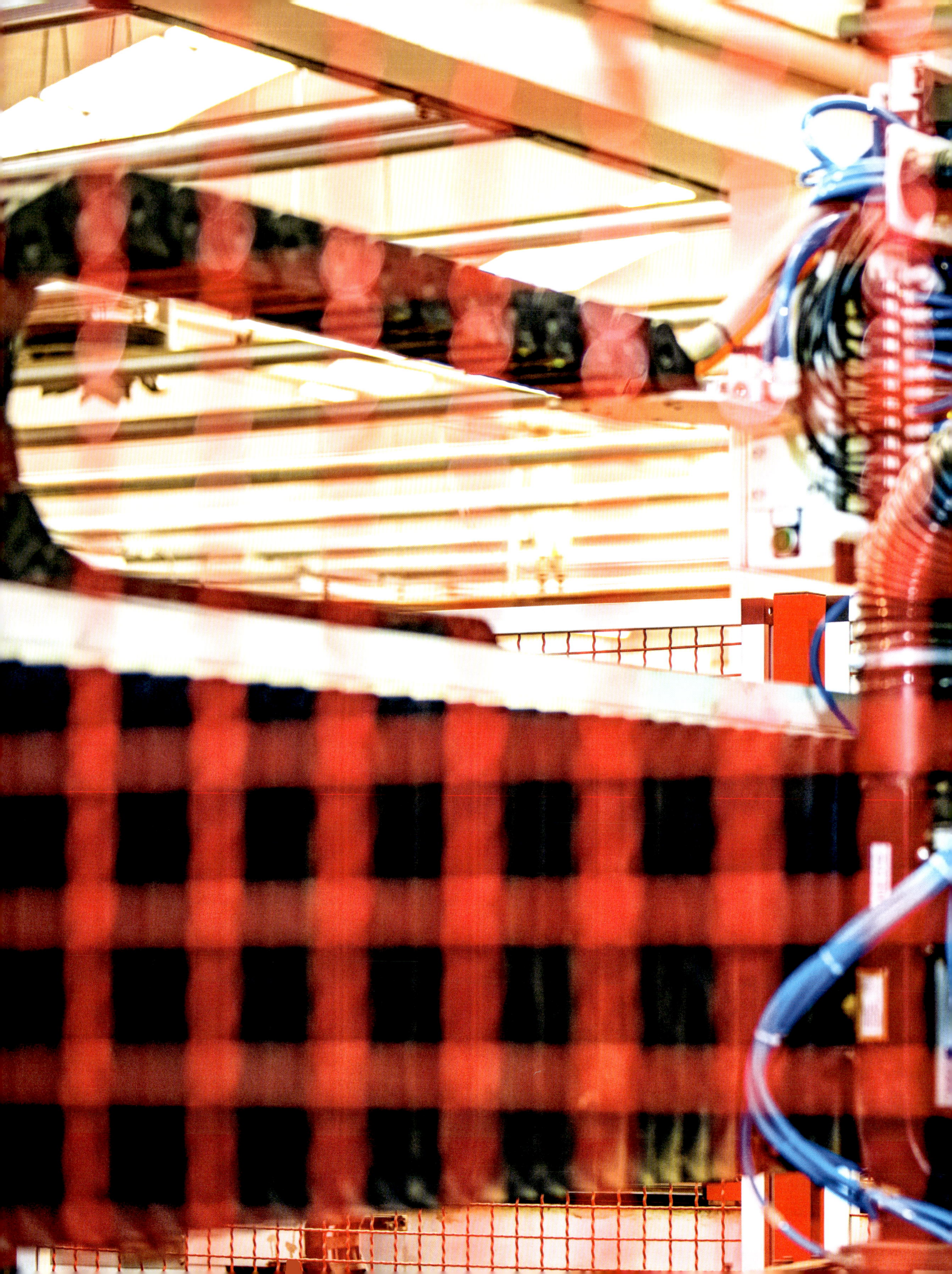

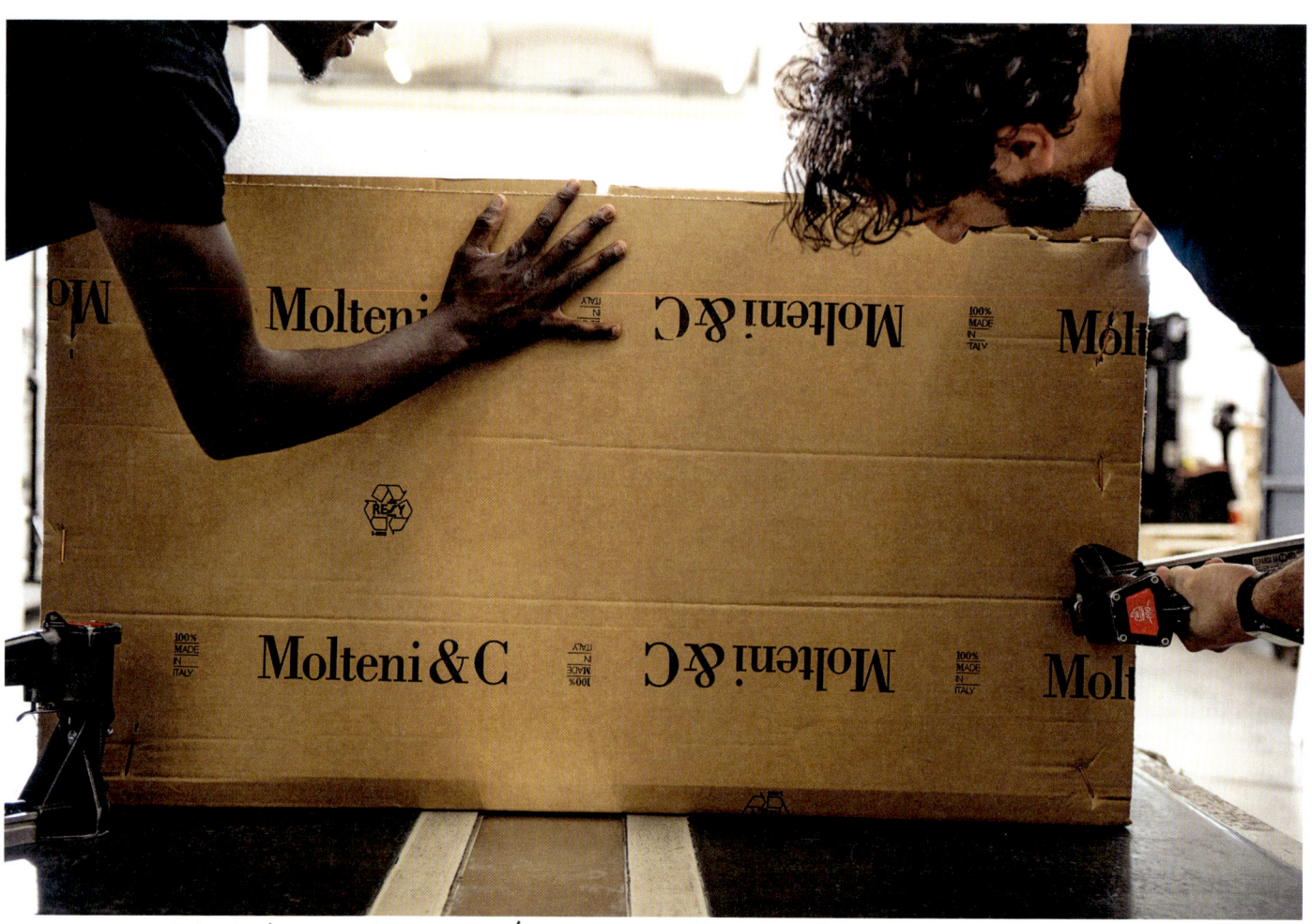

Finishing the packaging and readying the boxes for shipping

A stock of wardrobe pieces awaits shipping

96734 A

96724 A

96718 A

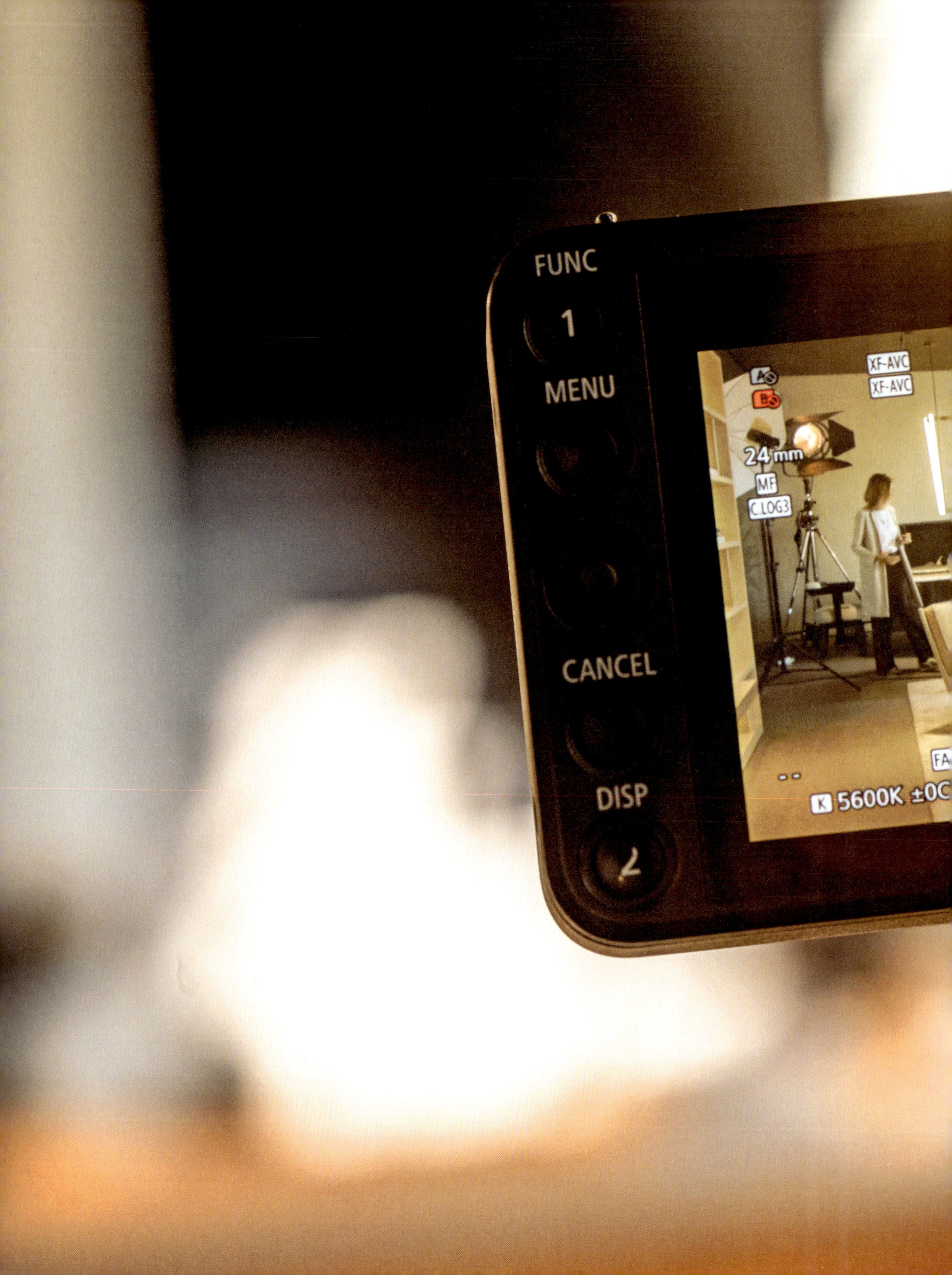
FUNC
1
MENU
CANCEL
DISP
2
XF-AVC
XF-AVC
24 mm
MF
C.LOG3
5600K ±0C

FRANCESCA MOLTENI

Inside the three-floor SHOWROOM at the Molteni&C Compound, which showcases the latest pieces from the collection and reproduces the ambience of a Molteni&C home

The space is used for training, film shoots, and interviews

GIULIA MOLTENI

Answering a journalist's questions...

A recessed handle expressing Molteni&C's high level of craft

The stovetop of the INTERSECTION kitchen

An open technical column of the INTERSECTION kitchen

Thermo oak shelves with integrated LED lights in the RATIO kitchen (2019), designed by Vincent Van Duysen

The sink of the RATIO kitchen

The RATIO kitchen

A detailed view of the HIGH-LINE 6-frame-door kitchen (2018)

HIGH-LINE 6-frame-door kitchen

"ULTIMATELY, [...] BEAUTIFUL COMPA[NY...] TELL IT'S ALL [...] IN ITALY. THERE [...] HERITAGE AND L[...] TAKING V[...]

"MOLTENI&C IS A
COMPANY WHERE YOU CAN
SEE CRAFTED AND MADE
IT IS AN INCREDIBLE
LEGACY THAT WE'RE
WITH US."

VINCENT VAN DUYSEN

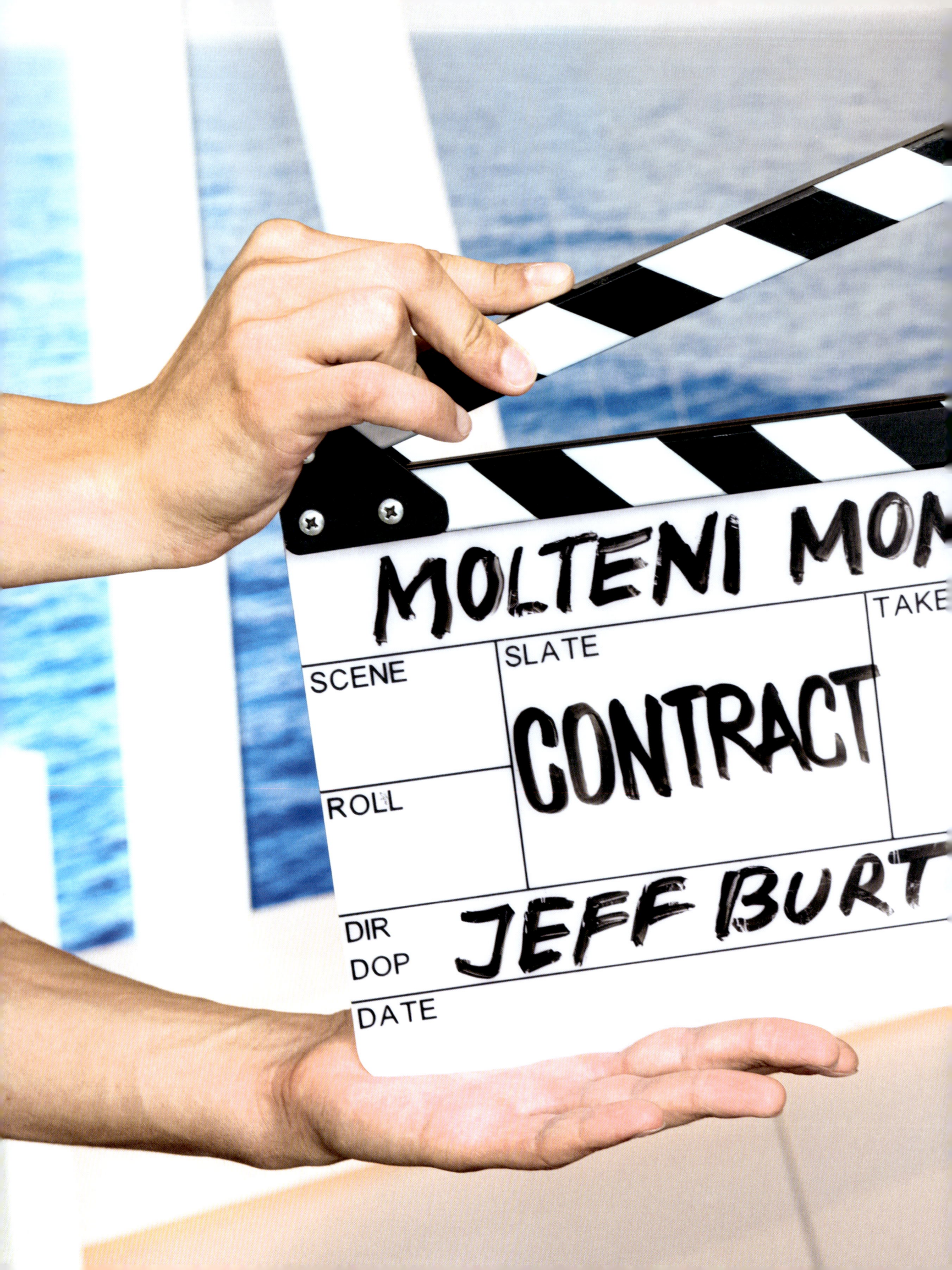

CONTRACT DIVISION

By Ali Morris

Since its launch in 1975, Molteni&C's Contract Division has collaborated with some of the most respected names in architecture and design to produce more than 500 tailor-made projects across the globe. In the rich cultural context of these prestigious public spaces and buildings, Molteni&C's design and production prowess comes to the fore. Among the group's collaborators are no fewer than six Pritzker Prize-winning architects, including Norman Foster, Jacques Herzog and Pierre de Meuron, Jean Nouvel, Renzo Piano, Aldo Rossi, and Alvaro Siza. For these industry titans, the draw to Molteni&C is undoubtedly in large part its Contract Division, which has unparalleled expertise and ability in bringing innovative and bespoke solutions to life, and often even serves as the creative force that leads to some of the products in the Molteni&C collection, most notably in the case of Jean Nouvel's Less table (1994), originally designed for the Fondation Cartier in Paris.

Made up of architects, engineers, and technicians with specialties in woodworking, metalworking, and upholstery, the Contract Division team works hand in hand with the group's dedicated production center and an extensive network of exclusive providers and subcontractors. This deep knowledge of materials, combined with a commitment to research, experimentation, and innovation, enables them to devise custom pieces that exemplify Molteni&C's nearly century-long obsession with quality and detail.

At Renzo Piano's New York Times Building in Manhattan, which opened in 2007, Molteni&C's office design brand, UniFor, was called upon to create a rational, flexible, and non-hierarchical workplace. Beyond furnishing a staggering 2,480 open-space workstations, 300 closed offices, 410 conference rooms, and 850 wall storage and filing cabinets, UniFor also created custom-made furnishings for task areas and offices. At Herzog & de Meuron's steel and glass Feltrinelli Foundation, UniFor provided the furnishings and partition systems for the three upper floors. Opened in Milan in 2016, the commission included the top-floor reading room, which was furnished exclusively with custom works designed by the architects to complement the forms and geometries of the building. In another collaboration with the Swiss architects, 56 Leonard Street, a 60-story residential tower in New York's Tribeca neighborhood that was completed in 2017, Molteni&C produced 146 custom kitchens featuring rounded black lacquer islands with sliding granite tops that nod to the tower's distinctive stacked glass form. The brand also contributed custom mirrored linen closets and 286 cabinets with mirrored sliding doors intended to reflect the spectacular city skyline. Other notable residential projects include John Pawson's Residences at The EDITION West Hollywood, Renzo Piano's 565 Broome SoHo, and Jean Nouvel's 53 West 53 tower.

    Highlights from the Division's extensive portfolio of hospitality projects from around the world include the residences at Richard Meier's Four Seasons Hotel at The Surf Club in Miami; the Mandarin Oriental Palace in Luzern, Switzerland; and the Borgo Santandrea, a cliffside hotel on the Amalfi Coast by the architect Rino Gambardella. Opened in 2022, the latter features re-editions of classic Gio Ponti designs produced by Molteni&C, such as the D.153.1, D.151.4, and D.156.3 armchairs and the D.555.2 small tables.

    Since 1999, Molteni&C has also applied its know-how to the creation of marine projects such as Cunard's MS *Queen Victoria*, an elaborate cruise ship that includes the fit-out of a three-floor theater with private boxes and numerous entertainment areas with the casino, dance floors, and the Queen's Room inspired by the former British royal residence Osborne House.

Cunard's **QUEEN VICTORIA**

The P&O BRITANNIA, an example of a Molteni&C contract project

Above, opposite, and following spread: Mock-ups of the SUN PRINCESS cruise ship, furnished by Molteni&C's contract division

H.6310 - AREA II SUITE RESTAURANT dk

LOS ANGELES Flagship Store, 2019

"Aldo Rossi: Design 1960-1997" exhibition at Museo del Novecento in Milan in 2022

GRAN TEATRO LA FENICE (2003) by Aldo Rossi, with executive design by Arassociati

The monomaterial INTERSECTION kitchen (2020), designed by Vincent Van Duysen

Inlaid top of the kitchen

GIO PONTI COLLECTION

Remade since 2012 in partnership with the Gio Ponti Archives and under the artistic direction of Studio Cerri & Associati, the Gio Ponti Collection from Molteni&C has not only provided an opportunity to unearth and rediscover many of Ponti's forgotten designs, it's also been a key way to begin again from these particular origins of modernity. Though in many ways the project is about bringing traces of the past into the present, it has turned out to be an exercise that's decidedly contemporary in feeling, with fresh directions to be explored and artisan know-how to be interpreted.

Reproduced on an industrial scale and in numbered editions with Ponti's signature, each complete with a certificate of authenticity, the Gio Ponti Collection pays tribute to one of the most complex architects of the twentieth century and serves as a rare opportunity to renew attention for this leading figure of architecture. In a career spanning more than 50 years, Gio Ponti (1891-1979) founded *Domus* magazine, lectured at the Polytechnic University of Milan, and designed everything from buildings and interiors to objects and furniture—including designs that, without Molteni&C's intervention, could very well have been lost to time. Ponti completed more than 100 buildings in his lifetime, among them the Pirelli Tower in Milan (1960); the Hotel Parco dei Principi in Sorrento, Italy (1962); and the Denver Art Museum in Colorado (1971).

The lengthy process of researching, studying, and selecting—conducted in the archives with Ponti's heirs, in particular with the curator Salvatore Licitra — led to remakes of historic items made by the great designer for private homes and special projects, or in small-edition series. While respecting the originals, these additions are produced industrially, applying the latest technologies to bring them up to date and in line with Molteni&C's *"total living"* concept. Enriched with new introductions each year, the Collection includes furniture that Ponti designed between 1935 and the '50s, such as a bookcase, a chest of drawers, a small table, a frame, and a rug for Casa Ponti on Via Dezza in Milan (1957); an armchair for Villa Planchart in Caracas (1953-57); and a table for the Time & Life Building in New York (1959).

In 2017, Molteni&C signed a 10-year exclusive worldwide licensing contract for producing the Collection, further deepening its relations and trust with Ponti's heirs. *"There are fascinating links between industry and art,"* Ponti once said. Fittingly, the Gio Ponti Collection is a celebration of these links.

Gio Ponti books in the archive

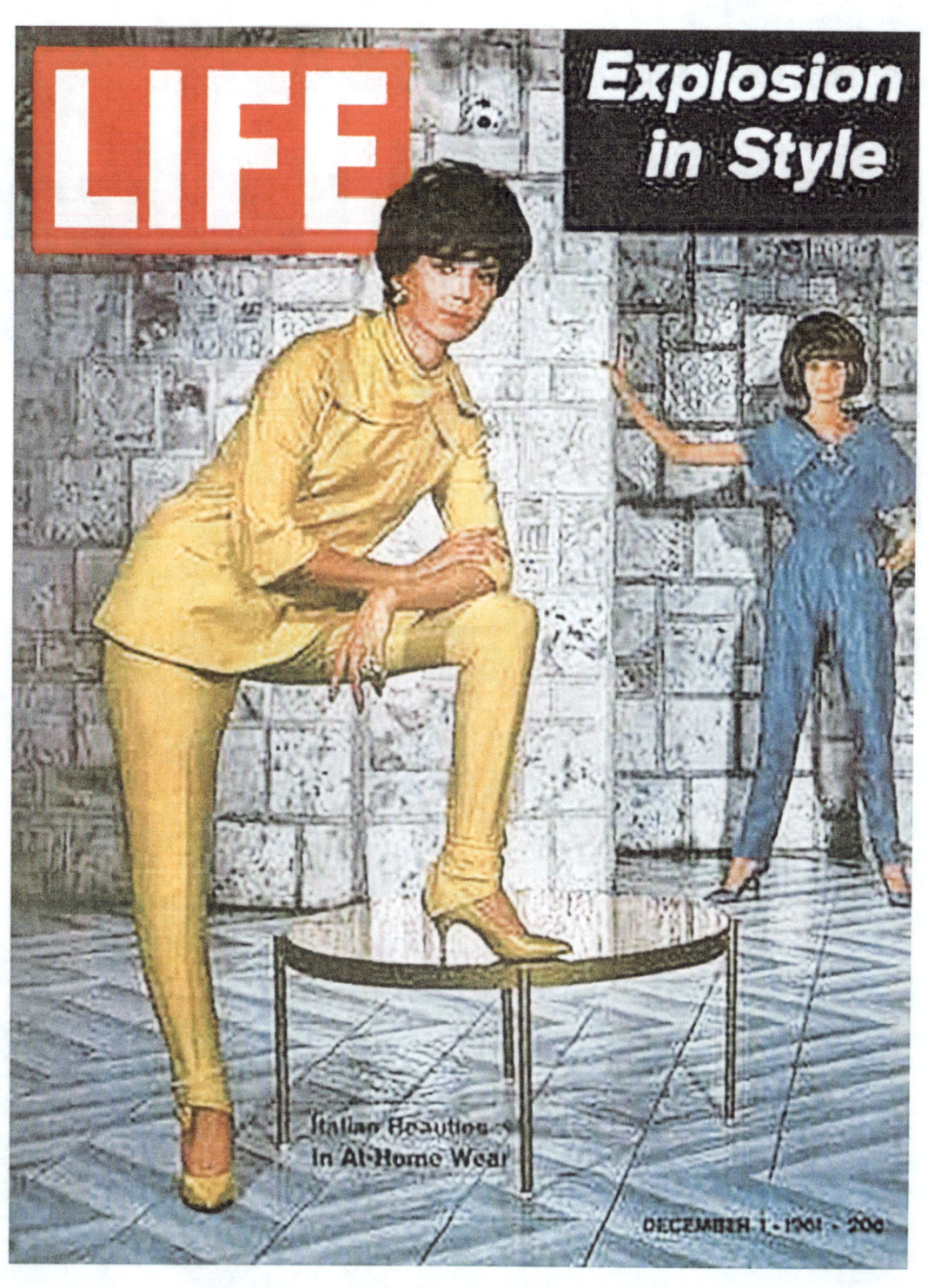

A cover image of *Life* magazine, photographed in Alitalia's Fifth Avenue office in New York City, designed by Ponti in 1958

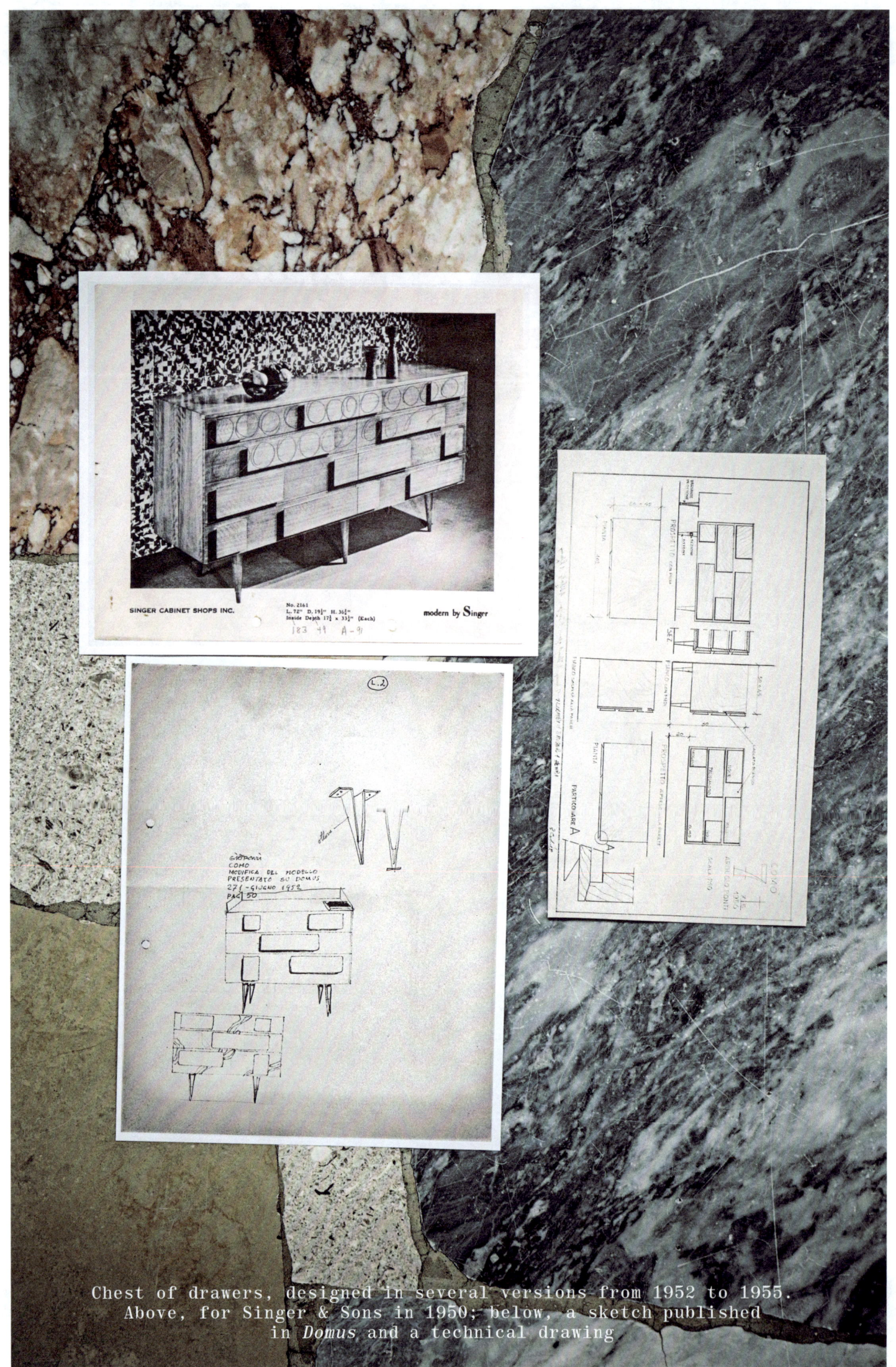

Chest of drawers, designed in several versions from 1952 to 1955.
Above, for Singer & Sons in 1950; below, a sketch published
in *Domus* and a technical drawing

The D.655.1 model by Molteni&C

Top to bottom: The Liberty Stores exhibition in London (1957).
Ponti's D.552.2 table

One of the bedrooms of the Ponti family house on Via Dezza in Milan

Armchair designed for Ponti's Villa Planchart in Caracas (1953-57), here shown at the XI Triennale in Milan in 1957

Top to bottom: The D.859.1 table,
designed for the Time Life Building in New York City (1959).
The Round chair, or Otto Pezzi ("*Eight Pieces*"), from 1956

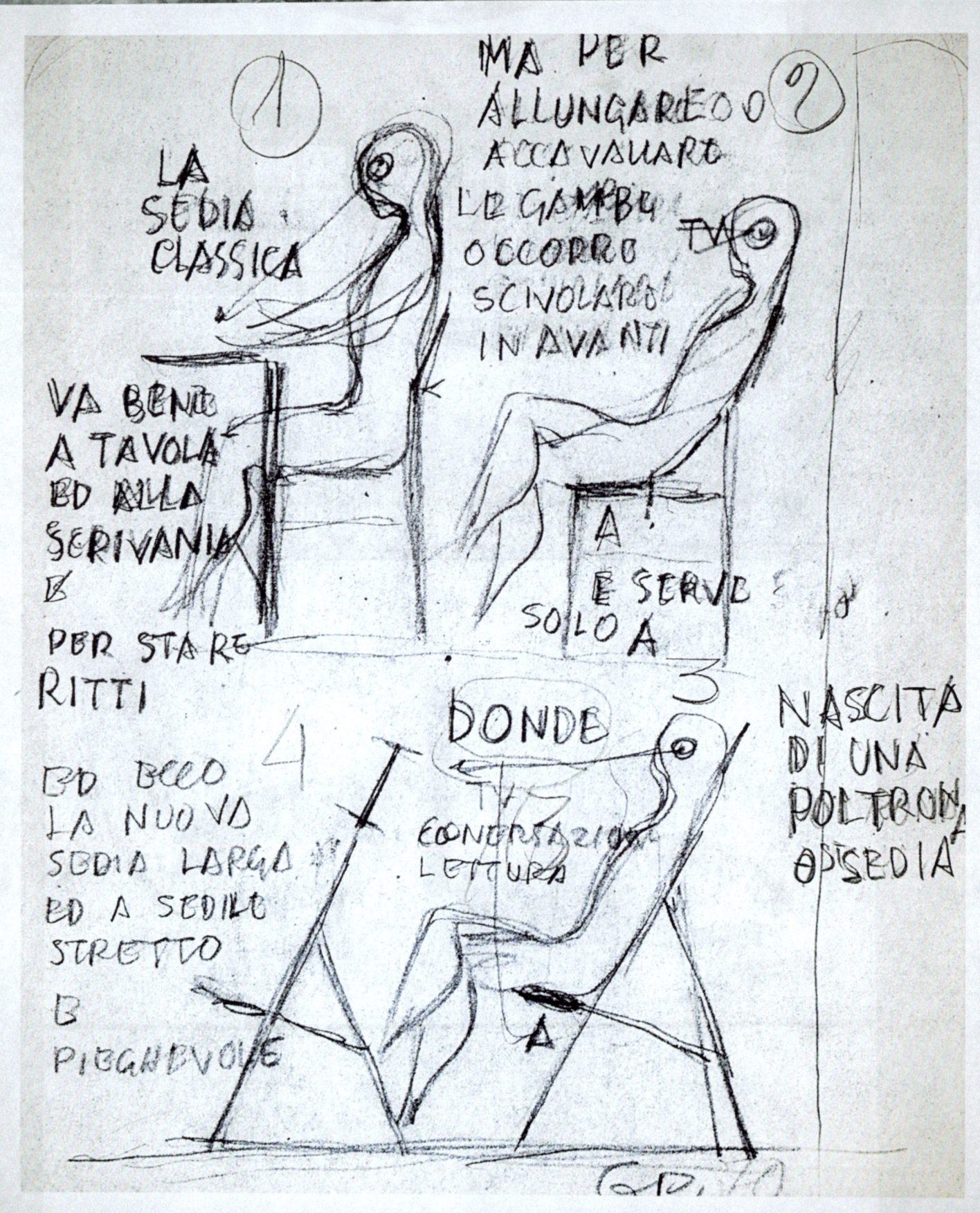

Drawing by Ponti exploring different sitting positions and chairs

Chair from the Apta collection for Walter Ponti,
presented at the Eurodomus 3 exhibition in Milan in 1970

Prototypes of Ponti's *"Painted Doors"* and *"Printed Doors"* with Jsa fabrics (1955-57)

Ponti's D.754.1 carpet in cavallino produced by Colombi for Altamira in 1954 and reissued by Molteni&C as part of the Heritage Collection

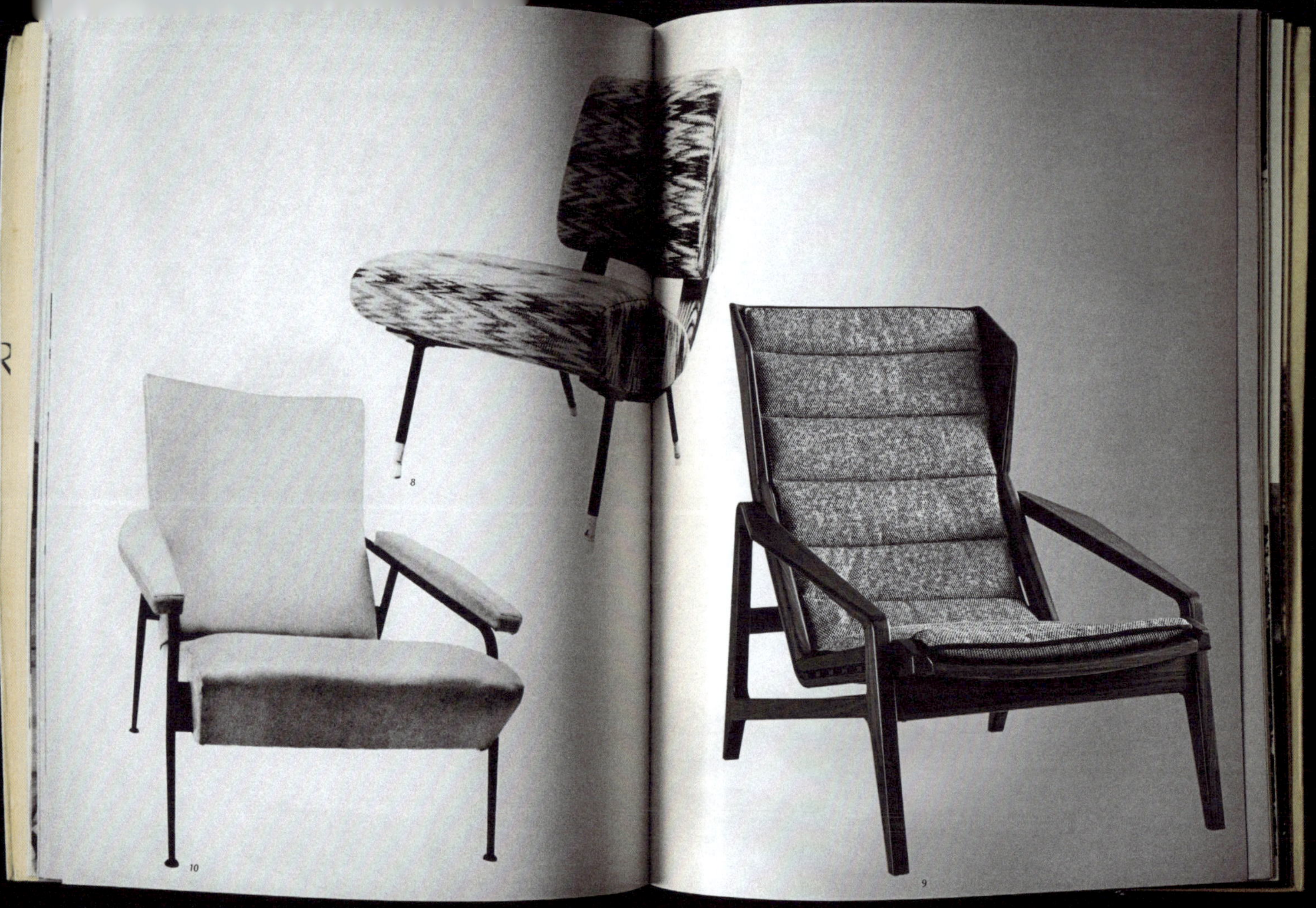
8
10
9

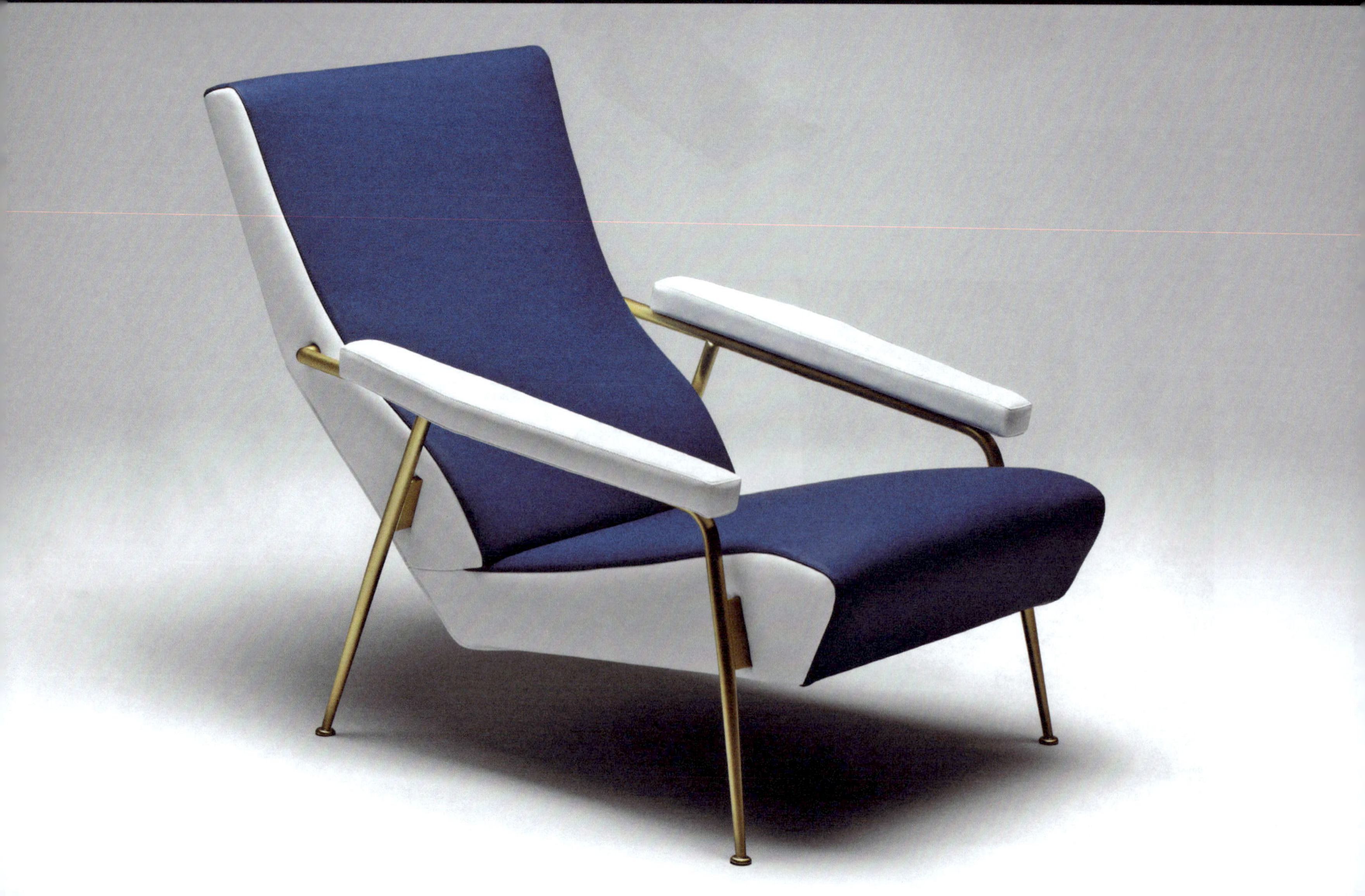

A presentation of Ponti designs at the XVI Mostra internazionale di Barcellona in 1948

Opposite (top): A magazine spread showing, from left, the Distex armchair (1953), the Round armchair (1956), and the D.156.3 armchair (1956), now reissued by Molteni&C and originally produced for the American company Altamira. Opposite (bottom): The D.153.1 armchair (1953), originally designed by Ponti for his house and reissued by Molteni&C

The chair reissued by Molteni&C

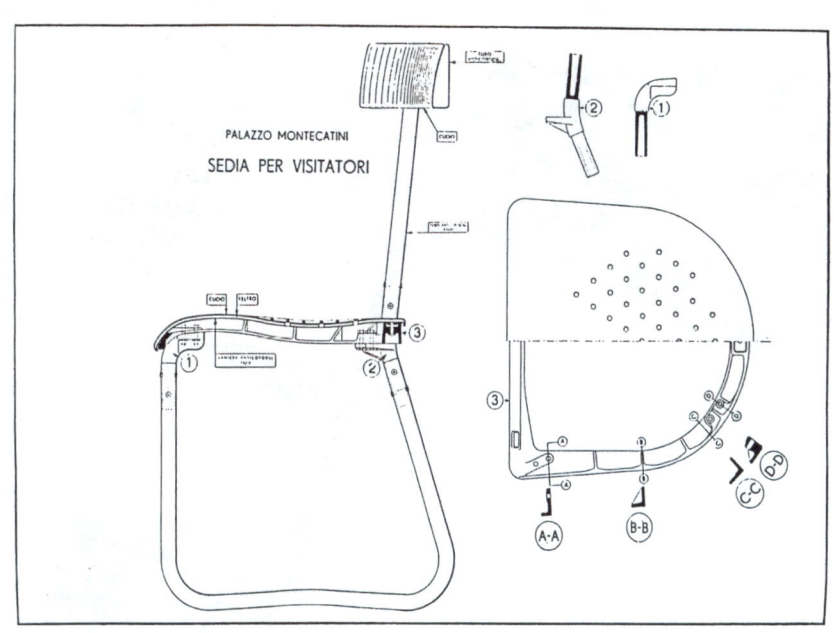

Picture of Palazzo Montecatini in Milan (1935)
and drawing of the aluminum chair

Ponti's D.555.1 coffee table (1955), reissued by Molteni&C

A bespoke jewelry display case, created for a renowned French jewelry brand

A series of bespoke jewelry display cases

The FONDATION CARTIER in Paris featuring the LESS table, both designed by Jean Nouvel for UniFor

565 BROOME SOHO, designed by Renzo Piano, in New York City

WEST 53, designed by Atelier Jean Nouvel, in New York City

56 LEONARD STREET, designed by Herzog & de Meuron, in New York City

AMABILA SUITES, Milan

MOLTENI&C'S C[...]
HAS PRODUCE[...]
TAILOR-MADE [...]
THE GLOBE WI[...]
MOST RESP[...]
IN ARCHITECTU[...]

NTRACT DIVISION
MORE THAN 500
ROJECTS ACROSS
SOME OF THE
CTED NAMES
RE AND DESIGN.

The <u>WEST HOLLYWOOD EDITION</u> hotel, designed by John Pawson, in Los Angeles

<u>FOUR SEASONS HOTEL</u> at the Surf Club, Surfside, Florida

BORGO SANT'ANDREA hotel in Amalfi, Italy

MANDARIN ORIENTAL PALACE in Luzern, Switzerland

BONNEFANTEN MUSEUM (1990-94) in Maastricht, The Netherlands, designed by Aldo Rossi with Umberto Barbieri, Giovanni da Pozzo, and Marc Kocher, and furnished by Unifor and Molteni&C

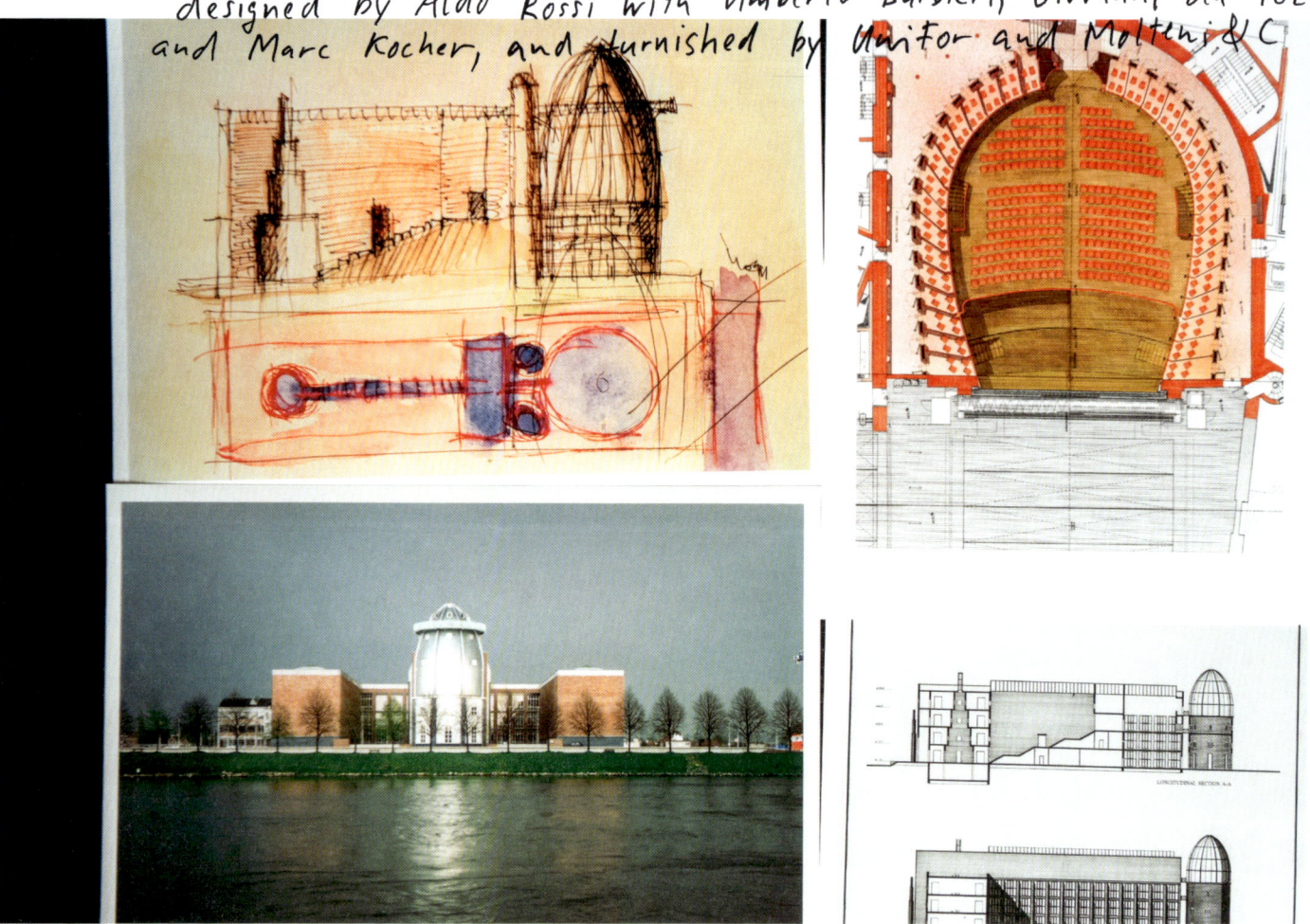

Top right: GRAN TEATRO LA FENICE (2003) by Aldo Rossi

The Salone del Mobile presentation of PIROSCAFO designed by Luca Meda and Aldo Rossi in the early 1990s

A selection of Molteni&C's Flagship stores around the world

TAIWAN

AMSTERDAM

CAPE TOWN

SEOUL

The Molteni Museum at the Molteni & C headquarters, curated by Ron Gilad

It presents key pieces that mark the history of the group

A moment of reflection inside the Molteni Museum.
Arc table by Foster + Partners (2010)

The museum comes alive during a vernissage

The museum serves as a place where the past, present, and future come together

The museum offers illuminating insights through archival images and objects

A "ROOMSCAPE" imagined by <u>Vincent Van Duysen</u>

enhancing the connection between indoor and outdoor, with pieces from the OUTDOOR collection

Another "ROOMSCAPE" as imagined by Ron Gilad

Conceived as a "theater of the absurd," the installation plays with perception and suggests an open-ended idea of space and objects

A 3D-printed model of RON GILAD with a Teatro chair by Luca Meda and Aldo Rossi

All of the art pieces were produced by Molteni &C for Gilad's solo exhibition at the Tel Aviv Museum of Art in 2013

RODOLFO DORDONI
Chelsea Armchair
2015

ALDO ROSSI AND LUCA MEDA
Teatro Chair
1982

JEAN NOUVEL
Less Table
1994

LUCA MEDA
Vivette Armchair
1989

HANNES WETTSTEIN
Alfa Chair
2001

YASUHIKO ITOH
MHC.2 Bookcase
Re-edition, 2016
Heritage Collection, 1959

PATRICIA URQUIOLA
Diamond Table
2004

RODOLFO DORDONI
Filigree Table
2013

WERNER BLASER
MHC.1 Dresser
Re-edition, 2016
Heritage Collection, 1955

IGNAZIO GARDELLA
Blevio Table
Re-edition, 2022
Heritage Collection, 1930

GIO PONTI
D.235.1 Chair
Re-edition, 2012
Gio Ponti Collection, 1935-38

ALDO ROSSI
Milano Chair
1987

JASPER MORRISON
Tea Chair
2021

HERZOG & DE MEURON
Porta Volta Chair
2023

AFRA AND TOBIA SCARPA
MHC.3 Miss Chair
Re-edition, 2016
1986

FOSTER + PARTNERS
Teso Coffee Table
2013

ALDO ROSSI
Carteggio Single Unit
1987

RON GILAD
Panna Cotta Coffee Table
2012

LUCA MEDA AND ALDO ROSSI
Piroscafo Single Unit
1991

VINCENT VAN DUYSEN
Hector Bookcase
2020

MONUMENTI

By Janelle Zara

The geometries of Michael Anastassiades's Half a Square table (2020) bear a deceptive simplicity, where minimalist parts come together through skillful construction. The name playfully refers to the top and legs' intricate triangular joinery that resembles the diagonally cut half of a square once assembled. Offered in materials including graphite oak, marble, and glass, the tabletop *"almost hovers over the four legs,"* as Anastassiades himself puts it, a beguiling detail he credits to Molteni&C's in-house technical precision. *"It's like some magic happens,"* he says.

Subtle design elements rendered in rich materials – and fashioned with an enduring commitment to timeless sophistication and technical rigor – are the hallmarks of Molteni&C. As an international phenomenon, Molteni&C's roots are deeply aligned with the protagonists of twentieth-century Italian design and culture, producing such classics as the 1930 Blevio table, a design by the late architect Ignazio Gardella for his own family home in Lake Como. The reissue of the 1986 Miss chair revives early collaborators Tobia and Afra Scarpa's sinuous craftsmanship, where the back and legs appear as one harmonious line carved from wood. The great Luca Meda left his lasting impression on the brand, too, by designing icons like the geometric Vivette armchair and by bringing fellow legend Aldo Rossi into the company in the 1980s. Rossi's nearly 20-year partnership with Molteni&C produced the beloved Milano chair and Carteggio dresser (both 1987), as well as his and Meda's 1982 Teatro chair for Genoa's renowned Teatro Carlo Felice.

Embracing the industrial advances of the 1950s and '60s, the brand's midcentury evolution from artisanal to mass production appears almost effortless. Heightened technology did not erase tradition, but instead furthered Molteni&C's dedication to quietly robust silhouettes. Swiss architect Werner Blaser's 1955 chest of drawers and Yasuhiko Itoh's avant-garde 1959 walnut bookcase mark the brand's early forays into modern design, both winning prizes in Cantù's inaugural

Concorso Internazionale del Mobile. New materials and processes afforded unprecedented lightness, as in the automotive resin in Hannes Wettstein's 2001 Alfa chair, the light elastic surfaces stretched over Patricia Urquiola's 2005 Glove chair, and the thin origami-like legs of her 2004 Diamond table, or the base of Foster + Partners' Teso table (2013), a helix sculpted by a robotic arm. Jean Nouvel's 1994 table for Fondation Cartier, fittingly called Less, simply comprises a folded sheet of steel.

 Contemporary designers have continued Molteni&C's heritage of elegant simplicity, as with Rodolfo Dordoni's Chelsea armchair (2015) and seamlessly extendable Filigree table (2013), as well as Herzog & de Meuron's angular Porta Volta chair (2023). Ron Gilad's 2012 Panna Cotta coffee table playfully combines the weight of marble with fine iron legs, and a slight wave on the surface is reminiscent of its namesake classical Italian dessert.

 The brand continues onward under the creative direction of Vincent Van Duysen, whose modular Hector bookshelves (2020) adapted the modernist library into an inviting domestic system. The through line among all of these distinct pieces is their hidden qualities, where timelessness relies on the technical and functional features that are present, though often invisible. As Jasper Morrison, the designer of the Tea chair (2021), once said: *"Design in objects should be sensed rather than seen."*

The Molteni family goes through pictures showing the history of the group

HISTORY

By Emma Leigh Macdonald

On the occasion of Molteni&C's eightieth anniversary, celebrated in 2015, a catalog and exhibition curated by the designer Jasper Morrison emerged: *"80!Molteni."* Now, as the company turns 90(!), the exclamation point remains a fitting way to punctuate a design company that has had such a vast and wide-ranging influence on the design industry. While continuing to honor its storied legacy, the company stands at a forward-looking juncture, as described by members of the Molteni family running the company today, perhaps making the combined exclamation point and question mark, the interrobang, even more fitting for this year's celebration and for those to come: 90? 100? This punctuation mark was designed in 1962 by the advertising executive Martin K. Speckter, who introduced it to the United States through an article titled *"Making a New Point, or, How About That."* Coincidentally enough, at this juncture in time, across the Atlantic, Molteni&C was already nearly three decades into operating its business, one that continuously defined itself as *"making a new point."* And how about that?

When Molteni&C was founded in 1934 by the young couple Angelo and Giuseppina Molteni, it began as a craft workshop intended to continue the tradition of carpentry and upholstery of Giussano, Italy, and neighboring Brianza, which is particularly known for its history of woodwork design. Molteni&C's expansion into kitchen design with Molteni&C Kitchen's Dada Engineered in 1979 marked one of the company's most significant moments of growth and its first major diversification, which led to office design next, under the UniFor and Citterio umbrellas, and later to outdoor furniture collections. Navigating the world of interiors post-Covid-19, the company has proven itself adaptable to what the workspace means today, with flexible designs that will only continue to evolve with society's culture and needs.

Perhaps the company's reverence for craft and history on the one hand, and adaptability on the other, is thanks in part to the way it got its start: reproducing classic furniture styles in the traditional craft of the region. Their aesthetics are not often linked, but considering where Molteni&C began and where it stands today brings to mind a much larger-scale version of what the practice of the architect and furniture designer Eileen Grey (1878-1976), famous for her E-1027 villa in the Côte d'Azur, could have had: modular designs based in material-led craft that are timeless in their simplicity yet adaptable as aesthetic preferences and lifestyles evolve – the crux of designing successfully for how people actually live.

    While the company today is composed of over 1,200 employees across more than 100 countries, it continues to be family-oriented in a way that is coherent with its Italian context and the company's enduring values, but is also exceedingly rare in an industry that has increasingly become conglomerated. The fact that, today, Carlo Molteni, the son of the original founders, sits as the president of the company, with his children, Giulia and Giovanni, and nephew Andrea also at the helm – is a testament to the tradition-backed approach that is, alongside its commitment to research and innovation, embedded in the company's DNA. As Andrea has stated, *"We were raised like this, and this is how all of us have chosen to live."* In a contemporary era that suffers from decreasing specialization in crafts and other creative professions, Molteni&C is an inspiring answer – a managerial beacon – to the question of what it means to be a family company today.

Angelo Molteni on his bicycle at the Molteni&C plant in 1972

The Workforce in 1947.
The business already numbered over 60 people

"Arredamenti di Angelo Molteni" was quickly developing
less of an artisan workshop and more of a factory

Giulia Molteni points to a young Carlo Molteni and his father, Angelo

Angelo Molteni (born 1912) with his wife, Giuseppina, in 1941

Angelo set up his own business in Giussano, Italy, today the furniture district of Monza and Brianza

His artisan workshop became synonymous with top-quality products

At the beginning of its history, Molteni & C was well-known for processing the entire production of its furniture, starting with the tree trunk

A cake celebrating the 90th anniversary of Molteni&C

THE STORY CONTINUES...

AFTERWORD

By Jacques Herzog

Molteni&C is a company that stands for quality and tradition. They have pieces in their collection by architects like Gio Ponti and Ignazio Gardella, whose work we respect. Italy – a country often associated with political, economic, and cultural upheaval – is surviving its crises pretty well, thanks in part to strong, globally relevant brands such as Molteni&C, Feltrinelli, and Prada, with which we have been working over many years. Each time, the collaboration is hands-on and real. Interestingly, all of these companies are still in the care of their founding families.

Design as an independent discipline never interested me. I never felt attracted by its products, whether it was chairs, lamps, or bookshelves. Quite the opposite – I felt a kind of antipathy. Cars and planes were also never on my radar. I cannot explain this, really. Maybe it has to do with the sense of perfection that is inherent in many design objects. Perhaps this is why I feel more attracted to art, where I often see more imperfection, more room to discover.

In its beginnings, Herzog & de Meuron was a lot about discovering, about imperfection. We were *"bricolage-ing,"* experimenting with technical tools like video, and doing things with our own hands. We had to find our path in a professional field that was dominated by postmodernism and looming deconstructivism – both of which we could not identify with. Among all of the materials we were using at that time, wood was the one that worked best because it was easy to get, and we had the necessary machines and tools to work with it. We also liked the fact that wood, as a material for architecture, was considered *"uncool"* in the mid-1970s.

There was a carpenter in our neighborhood who had a lathe for turning wood. Normally, he did small and often kitschy things for antique furniture that he had been asked to repair. But he could also do very large objects, like columns spanning the full height of my studio. That is how I started to produce objects in the first place, using this old technology.

It allowed us to do things with rather unlikely shapes many years before the digitally operated milling machines started to flood the aesthetic landscape of architecture and design. These early wood columns and installation pieces could be seen as a kind of precursor to the objects and furniture pieces that we're now doing since we launched H&dM Objects. H&dM Objects summarizes the activities in our workshop, where a team of professional engineers and craftsmen develop prototypes for all kinds of specific and eventually also unspecific uses – unspecific in the sense that they lack a predetermined function but could guide us toward something new and unexpected.

    Porta Volta is what we named a wood armchair industrially produced by Molteni&C after we developed it in our workshop. We did many prototypes before it was fit for industrially scaled production with all functional and safety standards fulfilled. Normally, furniture pieces such as stools, tables, and lamps are the result of our architectural projects, created within a concrete spatial or programmatic context. This armchair did not have such a context; rather, it was something I would use for my own purposes, for which I had never found an adequate solution to what I wanted. I was probably not clear enough about what I wanted, but one day I did a sketch that we thought should be developed farther. The sketch showed a chair made of four boards, all more or less the same size and U-shaped. As I said before, I am not familiar with a professional industrial design approach, but I guess that made it a real design idea, a conceptual guideline that I imagine most designers use when they start working.

    What I like about the Porta Volta chair is its proportion and width. Sitting in it is like being inside a space – you can move inside the chair. It doesn't orient you in one specific way, as many other chairs do. And I especially like the lounge version with its shorter legs, so it sits you lower. I generally like chairs and tables that are lower than standard height – closer to the floor, farther away from the ceiling. A different perception of the space that surrounds you. More grounded. This is especially rewarding when you sit in a group, discussing and exchanging between yourself and others.

    It happened that I felt uncomfortable with my chairs and tables at home. So I decided to simply shorten their legs with a saw. Design is often a bit too complicated, too pretentious, isn't it?

CAST
by Order of Appearance

VINCENT VAN DUYSEN
MIKE HOLLAND
PATRICIA URQUIOLA
JEAN NOUVEL
JACQUES HERZOG
MARTA FERRI
RODOLFO DORDONI
ALESSIO ROSCINI
MARCO MATURO
JASPER MORRISON
GLENN PUSHELBERG
GEORGE YABU
MICHELE DE LUCCHI
TOBIA SCARPA
MICHAEL ANASTASSIADES
FRANCESCO MEDA
NAOTO FUKASAWA
RON GILAD
GIO PONTI

Molteni&C

CONTRIBUTORS

BEDA ACHERMANN

is the founder of Studio Achermann, an international creative agency based in Zurich. A pioneer in lifestyle journalism, he gained significant recognition in the 1980s as the creative director of German *Men's Vogue*. Achermann has collaborated with renowned artists such as Andy Warhol, Helmut Newton, and Jack Pierson, as well as with designers and architects like India Mahdavi and Vincent Van Duysen. Throughout his career, he has served as the art director of many renowned books, including *François Halard: A Visual Diary* (Rizzoli, 2019) and *François Berthoud: Fashion, Fetish, and Fantasy* (Rizzoli, 2022), and worked with photographers such as Peter Lindbergh, Steven Meisel, Herb Ritts, Paolo Roversi, Mario Testino, and Bruce Weber, among others. Achermann has created advertising campaigns and produced books and magazines for prominent brands and institutions such as Cartier, Chanel, Ferrari, LUMA Arles, Swarovski, and Zumtobel.

SPENCER BAILEY

is a New York–based writer, editor, and journalist. The cofounder and editor-in-chief of the media company The Slowdown, he is the host of the Time Sensitive podcast. He has written for publications such as *Bloomberg Businessweek*, *Fortune*, *The New York Times Magazine*, and *Town & Country*, and from 2013 to 2018, he was editor-in-chief of *Surface* magazine. Bailey is the author of several books, including *In Memory Of: Designing Contemporary Memorials* (Phaidon, 2020), and co-chair of the board of trustees of the Noguchi Museum in Long Island City, New York. Through The Slowdown, he has worked with clients such as Grand Seiko, Flos, Google, The Leading Hotels of the World, and Van Cleef & Arpels.

JEFF BURTON

is a Los Angeles–based artist whose photography has been exhibited at the Guggenheim Museum in Bilbao and the Barbican Centre in London. He has collaborated on projects with brands including Cartier, Dior, Louis Vuitton, Tom Ford, and Yves Saint Laurent, and contributed to publications such as *French Vogue*, *Domus*, *Vanity Fair*, *Fantastic Man*, *Numéro*, and *The New York Times*.

MARIA CRISTINA DIDERO

is a Milan-based independent design curator, consultant, and author. The curatorial director of the Design Miami fair, she has also curated numerous exhibitions for institutions around the world, including "Nendo: The Space in Between at Design" at Israel's Design Museum Holon and "Fun House" by Snarkitecture at the National Building Museum in Washington, D.C. Her writing has appeared in magazines such as *Apartamento*, *Domus*, *L'Officiel*, and *Vogue Italia*. She was editor-at-large of *Icon Design* from 2018 to 2020, and now serves as the Milan editor of *Wallpaper**.

JACQUES HERZOG

is the cofounder, with Pierre de Meuron, of the Basel-based architectural firm Herzog & de Meuron, established in 1978. He studied architecture at the Swiss Federal Institute of Technology Zurich with Aldo Rossi and Dolf Schnebli from 1970 to 1975. He has been visiting professor at Harvard University since 1989, and was professor at ETH Zürich and cofounder of ETH Studio Basel Contemporary City Institute from 1999 until 2018. Together with de Meuron, Herzog has been awarded the Pritzker Prize (2001), the RIBA Royal Gold Medal (2007), the Praemium Imperiale (2007), and the Mies Crown Hall Americas Prize (2014). In 2015, he cofounded the charitable foundation Jacques Herzog und Pierre de Meuron Kabinett, Basel.

SALVATORE LICITRA
is the curator of his grandfather Gio Ponti's archive. Since the 1990s, he has devoted himself to furthering the knowledge of Ponti's work, setting up the important Gio Ponti Archives database. As a curator of initiatives relating to Ponti, his best-known include the exhibition "Tutto Ponti: Gio Ponti Archi-Designer" at the Musée des Arts Décoratifs in Paris in 2018 and "Gio Ponti: Loving Architecture" at the MAXXI in Rome in 2019. He is also the author of the book *Gio Ponti* (Taschen, 2021).

EMMA LEIGH MACDONALD
is a New York-based writer, editor, and chef. She has curated, written, and edited work for the Istanbul Design Biennial, the New Museum, Public Records, Blue Hill at Stone Barns, and the New Art Dealers Alliance.

ALI MORRIS
is a London-based writer, editor, and consultant specializing in design, architecture, and interiors. She has contributed to publications such as *DAMN Magazine*, *Elle Decoration UK*, amd *Wallpaper**.

JEAN NOUVEL
is a French architect whose contextual approach and ability to infuse a genuine uniqueness into each project he undertakes has consistently yielded buildings that transform their environments and indelibly mark the cities in which they are built. These include the Lyon Opera House (1993), the Lucerne Culture and Congress Center (2000), the Philharmonie de Paris (2015), the Duo Towers in Paris (2022), and the Rosewood Tower in São Paolo (2022). In the field of design, Nouvel has developed several collections related to his architectural projects and in 1995 expanded his practice with the creation of the company Jean Nouvel Design. His works have won numerous awards, including the prestigious Pritzker Prize in 2008.

FRANCESCA PICCHI
is a Milan-based architect, independent curator, and journalist. She is professor of design history at the Istituto Superiore per le Industrie Artistiche in Florence, and for 16 years she was on the editorial staff of *Domus*. She is the coauthor, with Paola Antonelli and Giulio Castelli, of the book *La fabbrica del design* (Skira, 2007).

HARRIET THORPE
is a London-based author, journalist, and editor. She writes about architecture, urbanism, travel, contemporary art, and design, and contributes to publications including *Icon*, *Wallpaper**, and *The World of Interiors*.

JANELLE ZARA
is a Los Angeles-based freelance journalist whose work can be found in several publications beginning with "art" (*Artnet*, *Artforum*, *ARTnews*, and *The Art Newspaper*), as well as in *The Guardian*, *T: The New York Times Style Magazine*, and others. She is the author of the book *Becoming an Architect*.

IMPRINT

Concept and Creative Direction
BEDA ACHERMANN

Photographs: **JEFF BURTON**

Editorial Direction: **THE SLOWDOWN, NEW YORK**
Editor: **SPENCER BAILEY**
Associate Editor: **EMILY JIANG**
Producer: **RAMON BROZA**

Design & Art Direction
STUDIO ACHERMANN, ZURICH
YVES GERTEIS, ORLANDO BRUNNER,
CHLOÉ BRAUNSCHWEIGER, MARKUS BUCHER

Art Buying & Production
STUDIO ACHERMANN, ZURICH
ANNE BIEHLE, JANNINA STUKER,
VANESSA OBRECHT

Contributors:
SPENCER BAILEY
MARIA CRISTINA DIDERO
JACQUES HERZOG
EMMA LEIGH MACDONALD
ALI MORRIS
JEAN NOUVEL
FRANCESCA PICCHI
HARRIET THORPE
JANELLE ZARA

GIO PONTI
Photography: **OLIVER HELBIG**
Text: **SALVATORE LICITRA**

©2024 MONDADORI LIBRI S.P.A.
DISTRIBUTED IN ENGLISH THROUGHOUT THE WORLD BY
RIZZOLI INTERNATIONAL PUBLICATIONS INC.
300 PARK AVENUE SOUTH
NEW YORK, NY 10010, USA

ISBN: 978-8-8918-4162-9

2024 2025 2026 / 10 9 8 7 6 5 4 3 2 1

First edition: SEPTEMBER 2024

All rights reserved. No part of this publication
may be reproduced, stored in a retrieval
system, or transmitted in any form or by any
means, electronic, mechanical, photocopying,
recording, or otherwise, without prior consent
of the publishers.

**THIS VOLUME WAS PRINTED AT ERRESTAMPA S.R.L.
VIA PORTICO 27, ORIO AL SERIO, BERGAMO
PRINTED IN ITALY**

Visit us online:
Facebook.com/RIZZOLINEWYORK
X: @RIZZOLI_BOOKS
Instagram.com/RIZZOLIBOOKS
Pinterest.com/RIZZOLIBOOKS
Youtube.com/user/RIZZOLINY
Issuu.com/RIZZOLI

GIO PONTI

INSIDE THE
GIO PONTI ARCHIVES
BY SALVATORE LICITRA

These pages showcase images from the Gio Ponti Archives, located on Via Dezza in Milan, ephemera found in the chaotic jumble that can inevitably ensue during urgent searching. The images are paired with marble or stone, as if the photographs, drawings, and magazine covers were portals into the world of Gio Ponti, openings in his beloved surfaces.

The different kinds of marble glimpsed here are, in fact, those of the incredible, irregular floor put together by Ponti throughout the rooms of his studio. Made with marble quarried in the Tuscan town of Montecatini, it is a sort of outsize Palladian floor mosaic, conceived to act as a stage on which to arrange furniture and create domestic settings, in the same way that the ceramic tiles he studied as external decoration for the façades of his architecture were destined to form a bright and colorful urban theater.

But let's talk about the archive these materials come from. I am very familiar with it by now; I've spent a lot of time here over the years, and know only too well what happens when you examine this place closely. Looking carefully here and there, the observer will be able to see the archive as Gio Ponti once did: as a repository of ideas, of images that thrive in juxtaposition with one another. It is not a normal archive sorted into categories extraneous to the creative content of its topics. Rather, it is a vault of visual memories, grouped by works, to which Ponti resorted with inventive urgency to lend form to intuitions that he then integrated into his vast body of work.

Under a constantly roving eye, the images emerge in the present. The visual echo of a drawing, a sketch, or a note resonates with its neighbor. Like the keys on a piano, this invites an attempt at a symphony, a daring juxtaposition, a semantic repercussion. It is, in short, a training ground for the eye rather than a collection tidily archived for posterity.

Browsing these materials with the Molteni family, our aim was to find meaningful furnishings to create a collection that was an expression of Ponti's work. We were continually conscious of–and struck by–his harmonious weaving of references that mesh architecture, space, drawing, painting, proportion, color, thickness, void, and solidity. It suddenly appeared to us as indispensable that the curve of an armchair back must find balance with the profile of a bookcase, or that a table leg should act as a pivot for the perspective view of a window or as a counterweight for chairs lined up along a wall.

The Moltenis are objective people, all facts and figures, but this charming Ponti carousel of images and ideas suggests not a distraction but rather the conviction that, prompted by passion, they should add some exceptions to their rigorous rules.

Self-portrait by Gio Ponti (1918)

Drawing for a 1967 book of Ponti's poems, *Nuvole sono immagini*, edited by Vanni Scheiwiller

Ponti and his daughter Letizia

Illustration of *"GIENLICA"* (Gio Ponti, Enrico Bo, Lina Bo Bardi, and Carlo Pagani) for *Stile* magazine, from 1942

Console table designed in 1947 for Spartaco Brugnoli,
reissued by Molteni&C

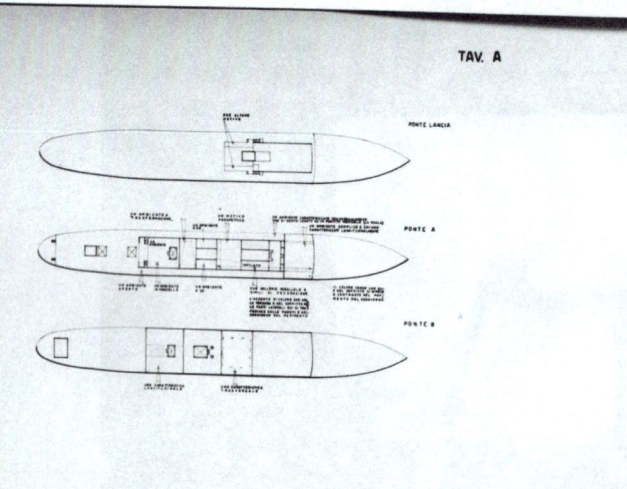

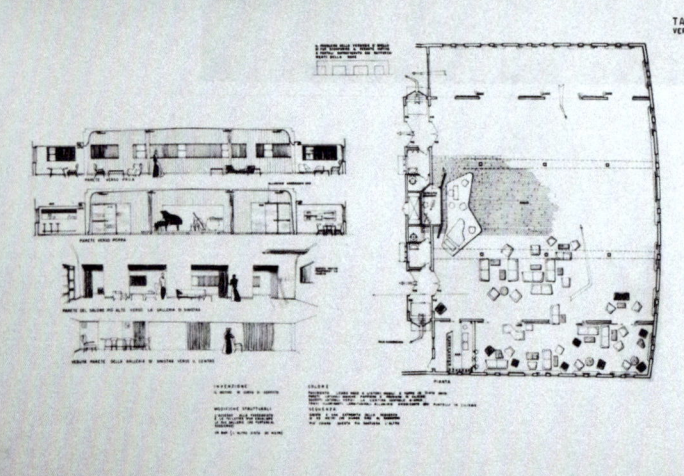

Drawings of the interior of the Conte Biancamano ocean liner
and the D.251.4 armchair (1949), reissued by Molteni&C

MURALS
by Gio Ponti

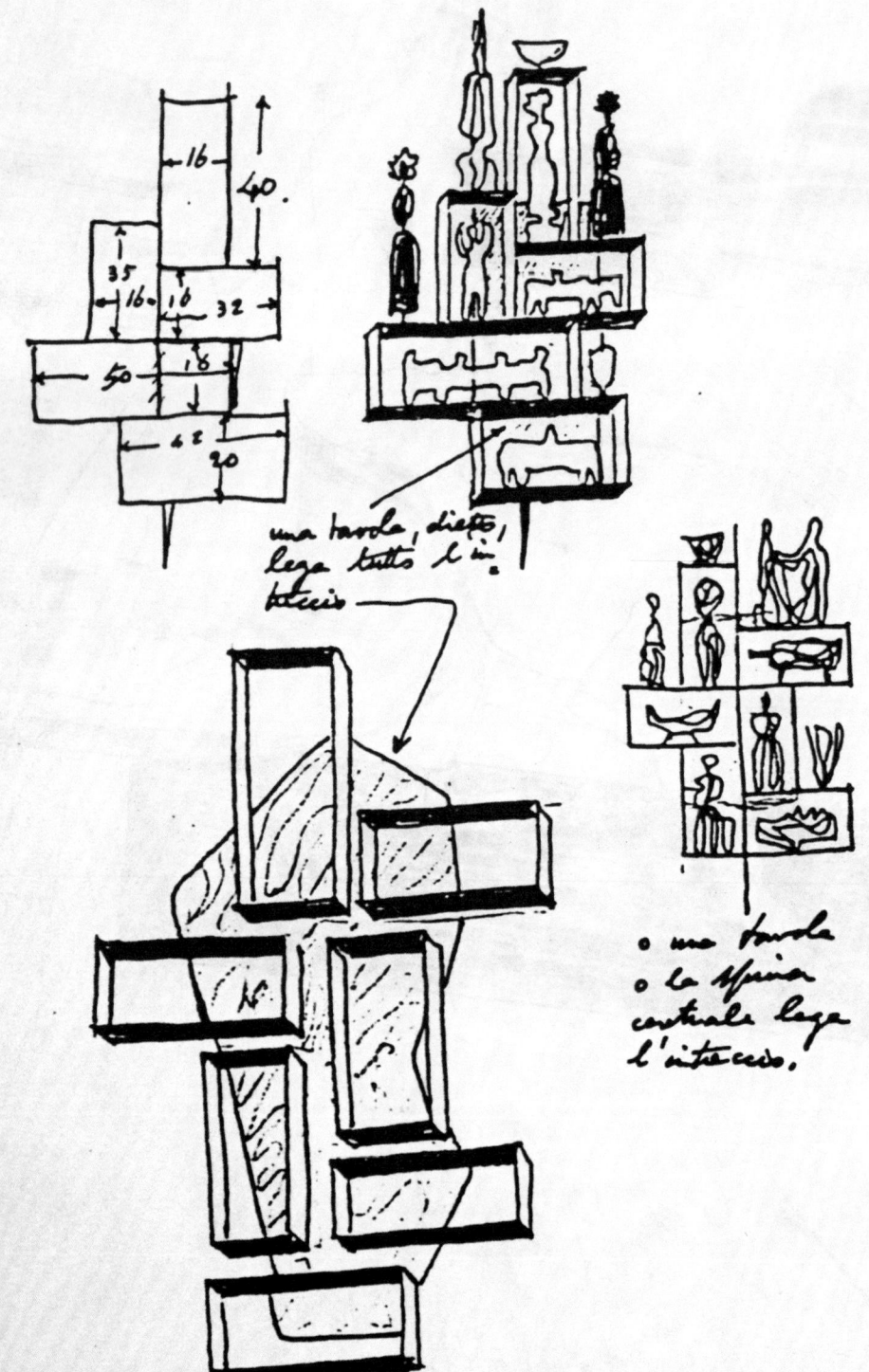

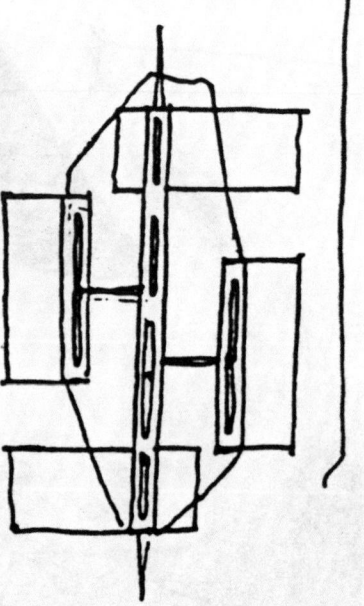

Drawings by Ponti

Bookcase designed by Ponti for his house in Via Dezza, Milan (1956-57), reissued by Molteni&C

Ponti's coffee table for M. Singer & Sons (1950)